Storm Spotting and Amateur Radio

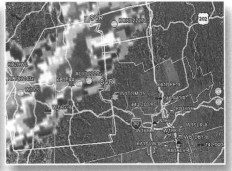

Authors
Michael Corey, W5MPC
Victor Morris, AH6WX

Editor
Steve Telenius-Lowe, 9M6DXX

Production
Mark Allgar, M1MPA

Cover
Sue Fagan, KB1OKW

Copyright © 2010 by

The American Radio Relay League, Inc.

Copyright secured under the Pan-American Convention

International Copyright secured

All rights reserved. No part of this work may be reproduced in any form except by written permission of the publisher. All rights of translation are reserved.

Printed in Canada

Quedan reservados todos los derechos

ISBN: 978-0-87259-090-8

First edition
First printing

Acknowledgements

A very heartfelt thanks to the members of the University of Mississippi Amateur Radio Club, W5UMS; Oxford-Lafayette County SKYWARN; the staff of the National Weather Service Memphis Field Office; Henry Leggett, WD4Q; Karl Bullock, WA5TMC; Malcolm Keown, W5XX; the American Radio Relay League, and all of those that contributed to this project.

This book is dedicated to my grandfather and elmer Fred Selley, K9MXG (SK).

Contents

Chapter 1	-	Introduction 5
Chapter 2	-	Safety 13
Chapter 3	-	Equipment & Resources 20
Chapter 4	-	Training 48
Chapter 5	-	Meteorology 55
Chapter 6	-	Hurricanes 76
Chapter 7	-	Storm Spotter Activation 92
Appendix 1	-	Weather Books for the Storm Spotter 115
Appendix 2	-	Weather Web Sites 116
		Index to SKYWARN Web Pages 117
Appendix 3	-	A Local SKYWARN Operations Manual 123
Appendix 4	-	Memorandum of Understanding between the National Weather Service and the American Radio Relay League, Inc. 125
Appendix 5	-	ARRL Reporting Forms 128
Appendix 6	-	False Statements notice 133
Appendix 7	-	Integrating *Google Earth*, NWS data and APRS using KML 134
Appendix 8	-	Sample After Action Report 139
Appendix 9	-	Lightning Protection for the Amateur Station 146
		(*QST* article published June - August 2002)
		INDEX 158

About the ARRL

The seed for Amateur Radio was planted in the 1890s, when Guglielmo Marconi began his experiments in wireless telegraphy. Soon he was joined by dozens, then hundreds, of others who were enthusiastic about sending and receiving messages through the air—some with a commercial interest, but others solely out of a love for this new communications medium. The United States government began licensing Amateur Radio operators in 1912.

By 1914, there were thousands of Amateur Radio operators—hams—in the United States. Hiram Percy Maxim, a leading Hartford, Connecticut inventor and industrialist, saw the need for an organization to band together this fledgling group of radio experimenters. In May 1914 he founded the American Radio Relay League (ARRL) to meet that need.

Today ARRL, with approximately 155,000 members, is the largest organization of radio amateurs in the United States. The ARRL is a not-for-profit organization that:
• promotes interest in Amateur Radio communications and experimentation
• represents US radio amateurs in legislative matters, and
• maintains fraternalism and a high standard of conduct among Amateur Radio operators.

At ARRL headquarters in the Hartford suburb of Newington, the staff helps serve the needs of members. ARRL is also International Secretariat for the International Amateur Radio Union, which is made up of similar societies in 150 countries around the world.

ARRL publishes the monthly journal *QST*, as well as newsletters and many publications covering all aspects of Amateur Radio. Its headquarters station, W1AW, transmits bulletins of interest to radio amateurs and Morse code practice sessions. The ARRL also coordinates an extensive field organization, which includes volunteers who provide technical information and other support services for radio amateurs as well as communications for public-service activities. In addition, ARRL represents US amateurs with the Federal Communications Commission and other government agencies in the US and abroad.

Membership in ARRL means much more than receiving *QST* each month. In addition to the services already described, ARRL offers membership services on a personal level, such as the Technical Information Service—where members can get answers by phone, email or the ARRL website, to all their technical and operating questions.

Full ARRL membership (available only to licensed radio amateurs) gives you a voice in how the affairs of the organization are governed. ARRL policy is set by a Board of Directors (one from each of 15 Divisions). Each year, one-third of the ARRL Board of Directors stands for election by the full members they represent. The day-to-day operation of ARRL HQ is managed by an Executive Vice President and his staff.

No matter what aspect of Amateur Radio attracts you, ARRL membership is relevant and important. There would be no Amateur Radio as we know it today were it not for the ARRL. We would be happy to welcome you as a member! (An Amateur Radio license is not required for Associate Membership.) For more information about ARRL and answers to any questions you may have about Amateur Radio, write or call:

ARRL—The national association for Amateur Radio
225 Main Street
Newington CT 06111-1494
Voice: 860-594-0200
   Fax: 860-594-0259
   E-mail: **hq@arrl.org**
   Internet: **www.arrl.org/**

Prospective new amateurs call (toll-free):
800-32-NEW HAM (800-326-3942)
You can also contact us via e-mail at **newham@arrl.org**
or check out *ARRLWeb* at **www.arrl.org/**

Chapter 1

Introduction

It is a Tuesday night in early February. The National Weather Service has issued a statement that there is a probability of severe weather coming into the area late that night. They have been tracking a squall line several hundred miles to the west. The storm has been producing strong winds and severe thunderstorms. Reports indicate that it has also produced hail and isolated tornadoes. A hazardous weather outlook is issued and reads "SPOTTER ACTIVATION MAY BE NEEDED".

Long before the storm arrives a local Amateur Radio SKYWARN® group is making preparations. Two stations volunteer to handle net control duties. Several members volunteer to go mobile if needed. Several more offer to relay weather reports from their home QTH. Phone calls are made to the non-Amateur Radio spotters in the area to keep them current on what is happening. Everyone checks to make sure all is in order; radios working, batteries charged, flashlights handy, go-kits ready, and vehicles fueled. The local Emergency Coordinator stays in touch with the National Weather Service and local emergency management, keeping up to date on the latest weather information. Now all that can be done is to wait.

February is not a typical severe weather month in this area. It is usually cool and rainy. But only a year before, almost to the day, a tornado outbreak struck the area killing 57 people and causing over a billion dollars in damages. The SKYWARN spotters were well aware of what was possible even in a normally slow severe weather month. As the spotters watch the line of storms approach they realize that this may be a very serious storm. Before it arrives the net control stations activate the SKYWARN Net. One is operating from the local Emergency Operations Center handling traffic coming in from the spotters. The other is relaying information to the SKYWARN Net one county to the east, keeping them abreast of the latest weather conditions. Mobile spotters are setting up to safely observe the storm as it approaches and being repositioned as needed. Home based spotters are calling in reports of conditions from their neighborhoods. Reports on the storm are being relayed to the National Weather Service office by radio, phone, and Internet. Reports are coming in of wind gusts at 60+ MPH, 1/4 inch diameter hail, street flooding, and a report of possible rotation in a cloud. Because of the incoming reports the National Weather Service can get a clearer picture of what is happening and can issue the appropriate watches and warnings.

The storm passes relatively quickly. In its wake come damage reports; trees down, power out, streets flooded, buildings damaged by wind. The spotter's job is not done. Calling in reports of damage is the next step. These after the event reports play a key part in understanding the weather. The local spotters, most also Amateur Radio Emergency Service (ARES) members, continue to submit reports. The local emergency manager has also asked that they assist with search and rescue efforts in a flooded area and help provide communications for a local shelter that has been set up.

This scenario is not new for Amateur Radio operators that serve as volunteer storm spotters. Amateur Radio has played a part in severe weather for decades. But long before SKYWARN, radar, the Internet, even before Amateur Radio, there was a critical need to get real time, ground truth information on severe weather that could be used to warn of approaching storms and aid in forecasting. To understand how we got to where we

Benjamin Rock, WX9TOR, storm spotting. (WX9TOR photo)

are today we should first look at severe weather in the US, the history behind severe weather observation, and the role communications have played.

HISTORY

Weather in the United States is about extremes. One of the earliest weather observations tells us this. In the 1600s William Bradford, governor of the Plymouth Colony, noted about winters in America "sharp and violent, and subject to cruel and fierce storms, dangerous to travel to known places, and much more to search an unknown coast."[1] Throughout American history there are accounts of deadly floods, tornadoes, thunderstorms, hurricanes, and wildfires.

Starting in the 18th century regular observations of weather conditions began to be made. The founding fathers were keen observers of weather. Thomas Jefferson made regular observations from 1772 - 1778 and used a thermometer and barometer. He even noted on July 4, 1776 the temperature in Philadelphia was 76°. George Washington kept daily weather observations in a diary until the day before his death. And Benjamin Franklin's weather experiments are part of popular American folklore.

In the 19th century a major advancement in communications, the telegraph, made it possible to relay weather observations to other stations. This was the starting point of weather forecasting. Advance notice of approaching weather had always been a need, but it was impossible without the ability to communicate real time observations. It was not long after the invention of the telegraph that organized observation of weather was started.

The Birth of a National Weather Service

In 1848 Professor Joseph Henry, Secretary of the Smithsonian Institution, proposed a program to observe and report severe weather. The program, The Smithsonian Institution Meteorological Project, would rely on a network of volunteers across the country. By the end of the first year the project had over 150 volunteers and by 1860 there were over 500. The system relied on the telegraph system to relay information from volunteer observers to the Smithsonian. In 1862 the Smithsonian published a pamphlet containing information about tornadoes and thunderstorms. The public was urged to take note of weather occurrences and forward the information on to the Smithsonian. As time went on other weather observation systems around the country, like that organized by Professor Cleveland Abbe in Cincinnati, would be absorbed into the Smithsonian's network.

During this time Professor Increase A Lapham, a scientist, author, and student of meteorology, urged that the federal government establish a weather service that could coordinate weather observations and issue forecasts. With the support of Colonel Albert Myer of the US Army Signal Corps he succeeded in lobbying Congress to pass legislation establishing a national weather service. In 1870 President Ulysses S Grant signed a bill that moved the weather service

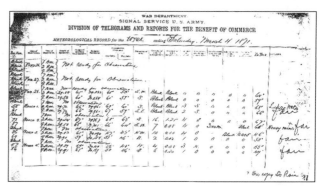

The first week of official observations taken in Memphis. (National Weather Service Memphis WFO)

from the Smithsonian to the Department of War. The new department was assigned to the Signal Corp and given the name The Division of Telegrams and Reports for the Benefit of Commerce. The use of volunteer observers was replaced with Observing-Sergeants who would telegraph weather reports to Washington DC. Initially there were 24 observing stations. By 1878 there were 284.

Along with the shift from the Smithsonian to the Signal Corps came the first organized weather training. There were very few professional meteorologists in the Signal Corps. The observation stations were made up of military personnel that had little or no background in weather spotting. At Fort Whipple in Virginia a course was developed within the curriculum of telegraph and signaling that covered meteorology and meteorological observation. Eventually a similar course was developed for officers. The training course lasted until 1886 when it was ended by the Secretary of War.

Due to internal turmoil the weather service was moved to the Department of Agriculture in 1890 by President Benjamin Harrison. The change created a new civilian weather bureau. The military personnel that had staffed the weather service under the Signal Corps were discharged and given the opportunity to go to the new department. With the transfer also came a name change: from 1891 - 1967 it would be known

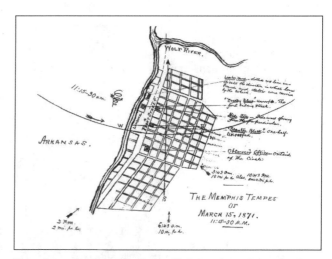

Map documenting the first official report of a tornado. (Memphis, Tennessee). (National Weather Service Memphis WFO)

as the Weather Bureau.

During this time monumental technological advancements greatly influenced how weather observations were reported and the ability to forecast weather. While many of these advancements were new meteorological technologies such as radar and weather satellite, the greatest advancements were in the areas of communication. In 1902 the Marconi Company began sending forecasts from the Weather Bureau to ships in the Cunard Line. And in 1905 the first weather observation from a ship at sea was sent by wireless. Wireless was a huge step forward over sending reports by telegraph. Telegraph required observation stations to be located along telegraph lines. With wireless an observation stations could be placed anywhere. As wireless advanced so to did the ability to receive real time weather observations. In 1921 the University of Wisconsin made the first transmission of weather forecasts by voice. By 1928 teletype had replaced voice for sending weather information.

Also during the early years of the Weather Bureau the use of volunteer observers reemerged. In 1890 the Cooperative Observer Program was established and is still in use today. The program has two goals; "to provide observational meteorological data, usually consisting of daily maximum and minimum temperatures, snowfall, and 24-hour precipitation totals, required to define the climate of the United States and to help measure long-term climate changes, and to provide observational meteorological data in near real-time to support forecast, warning and other public service programs of the National Weather Service."[2] In 1933 a science advisory group indicated to the President of the United States that the use of a volunteer cooperative observer network provided an extraordinary service. Today the program has more than 11,000 volunteers.

In 1940 the Weather Bureau was moved to another department within the federal government. President Franklin Roosevelt moved it to the Department of Commerce, where it remains today under the name National Weather Service, because of the role that it played in aviation and therefore the nation's commerce. During the years of World War Two the Weather Bureau, along with the military, set up a network of volunteer storm spotters. The initial mission of the spotters was to monitor lightning near ordnance facilities. The value of real time, first-hand information was apparent and by 1945 there were more than 200 observer networks in the United States. The use of weather observers was continued after the war. A series of tornadoes between 1947 and 1951 in Texas showed that having a local storm spotting network made it possible to get warnings out faster and save lives. Throughout the late 1940s and early 1950s spotter networks continued to develop. The use of spotting networks had a noticeable impact on the number of tornadoes being reported.[3] In May 1955 a tornado struck Udall, Kansas killing 80 people and injuring 273. This event prompted the Weather Bureau to recruit spotters and develop training for them. On March 8, 1959 the first official storm spotter training was conducted in Wellington, Kansas. 225 spotters received training. Along with the training the Weather Bureau issued a handbook for

The *Severe Storm Reporting Handbook* (1956) and *It Looks Like a Tornado..* (1959). (National Weather Service)

spotters, the *Severe Storm Reporting Handbook* (1956). This was followed up in 1959 with the publication *It Looks like a Tornado..*

In 1965 another weather event impacted storm spotting networks around the country. On April 11, 1965 - Palm Sunday - 47 tornadoes struck the states of Indiana, Illinois, Wisconsin, Ohio, and Michigan. This outbreak was the largest single day tornado outbreak in US history (since only passed by the super outbreak of 1974). Over 12 hours 271 people were killed and over 3500 injured.

Following the Palm Sunday tornado outbreak the Weather Bureau realized that there were serious shortcomings in early warning, communications, and storm spotting. From this came the National Disaster Warning System or NADWARN. A phase of NADWARN was the SKYWARN program. The SKYWARN program would be coordinated by the Weather Bureau (to become the National Weather Service in 1967) and train local volunteer storm spotters. Training classes were developed and the use of publications about severe weather was continued. SKYWARN started officially in the early 1970s. Since its beginning it has expanded to cover severe weather beyond tornadoes. Today there are over 280,000 SKYWARN trained storm spotters in the United States. The concept of SKYWARN has also made it to other countries with groups such as TORRO in the United Kingdom, SKYWARN Europe, and Canwarn in Canada.

Amateur Radio and Storm Spotting

Amateur Radio has played a part in responding to severe weather long before the development of SKYWARN, even before the development of organizations such as ARES and Radio Amateur Civil Emergency Service (RACES). Starting in the 1920s we see how Amateur Radio operators were involved when severe weather struck. During these early years the role was primarily in response to severe weather instead of ahead of it. Amateurs provided communications when normal means went down, relayed health and welfare traffic, and assisted various local, state and federal agencies during recovery. If we look at some of the stories from these early years we can see many similarities to what we do today.

In 1923 the Arkansas River, which crosses northeast Oklahoma, flooded. Normal lines of communication between

Tulsa and the near-by town of Sand Springs were lost. Amateur Radio operators Edward Austin, 5GA, Raymond McKinney, 5SG, and John Lewis, 5WX, of Tulsa and Halton Friend, 5XBF, and Howard Siegfried, 5GJ, of Sand Springs used Amateur Radio to maintain communication. The operators handled messages between the towns and relayed information from reporters to their newspapers. The operators continued this for three days when normal communication lines were restored.[4]

On January 8, 1937 an ice storm struck the central states of Missouri, Illinois, Arkansas, and Oklahoma. Across the region communication lines were down. Amateur Radio operators in the affected areas assisted by providing communications for almost a week before normal communication lines were restored. Traffic was handled for utilities such as electric and gas companies, telephone companies such as Southwestern Bell, the Associated Press, Western Union, and railroads. One station, W9PYF, logged 103 hours of operation, handing 183 messages.[5]

In September 1941 a hurricane struck the Texas Gulf coast. The hurricane, a category 3, known as Hurricane 2 (naming of hurricanes was not standardized until 1951) struck between Freeport and Port O'Conner on September 23. The storm caused a great deal of damage and took out communication lines in the areas around Galveston and Houston. An Amateur Radio operator, James 'Salty' Johnson, W5LS, of Houston, Texas volunteered to assist with communications. Before the storm made landfall he took a truck containing a 300W transmitter, generator, and receiver and went to the area where it was thought the center of the storm would strike. He waited the storm out in the city jail at Palacios, Texas. Salty was not alone, other Amateur Radio operators along the coast set up and prepared for the storm and to provide communications after it made landfall. Amateurs from around the state and around the country were ready to assist with disaster communications.[6] The death toll was only four. This was attributed to early warning and preparedness.

On the morning of September 29, 1938 a storm system moved into the Carolinas from the Gulf of Mexico. The storm produced five tornadoes that struck South Carolina. Two of the tornadoes were rated as F2 and three were rated as F1, resulting in 32 killed and around 100 injured. Assisting in the response to the devastation in Charleston, South Carolina were local Amateur Radio operators. As soon as power was restored to some areas amateurs went on the air and began handling traffic. Although telephone lines were not completely down they were overloaded by people calling to reach family and friends. Amateurs also assisted local law enforcement with communications for the police force and rescue efforts.[7]

Prior to World War Two most of Amateur Radio's involvement with severe weather was in response. The reason for this was the limited ability to forecast weather. Most Weather Bureau offices were still relying on little more than a thermometer, anemometer and wind vane, and barometer to detect changes in the weather. Innovations such as radar and weather satellites would not come until after the war.

> **W.E.R.S. EXPANDED**
>
> On April 10th some important changes were made by FCC in the regulations governing WERS, Part 15 of the FCC Rules, acting upon a request filed by ARRL about a year ago. As the regulations have read, WERS stations must be licensed primarily in connection with national security and defense and, although the regulations were changed some time ago to permit *such* stations to operate during natural-disaster emergencies, no provision has hitherto existed for the organization of WERS primarily on behalf of the relief of disasters. The regulations are now expanded to permit the CD type of WERS stations to operate in an emergency jeopardizing public safety, and new networks may be established and existing licenses renewed even though the Citizens' Defense Corps or equivalent civilian defense organizations are no longer active in the areas involved. WERS is the only agency through which amateur services may be utilized in the traditional duty of the amateur to cope with communication difficulties flowing from public emergencies, so this is a very significant expansion, as we comment upon in this month's editorial.

Article from the June 1945 *QST* on the War Emergency Radio Service.

The first effort at organizing any form of Amateur Radio network to do storm spotting came during World War Two. The War Emergency Radio Service (WERS) was started by the Federal Communications Commission in June of 1942. The regular Amateur Radio service was suspended during the war, but there was still a need for communications that Amateur Radio operators could fill. WERS was made up of licensed hams that operated WERS stations that were licensed to a community. The purpose was to assist with air raid warning and eventually expanded to include natural disasters and severe weather. Severe weather spotting was a secondary function that developed during the war. The original idea was to report on severe weather near military ordnance facilities. The weather observing component of WERS was formalized in 1945.

Following the war most Amateur Radio activity related to severe weather was still in providing assistance after the storm struck. Amateurs continued to set up emergency nets to relay traffic and assist with rescue efforts. Throughout the post-war years a combination of events changed the way Amateur Radio was involved in severe weather. Forecasting severe weather was advanced by the use of radar and weather satellite. The first tornado warning was issued, setting a precedent for future weather warnings. Amateur Radio technology was advancing to make it easier for stations to operate mobile and portable. Many of the storm spotter networks that started in World War Two were still active and new ones were being formed. Storm spotter training was beginning to take off. When SKYWARN finally came into being it was a perfect match with Amateur Radio.

STORM SPOTTING TODAY

Since the start of the SKYWARN program storm spotting has continued to evolve and advance. Originally the primary concern for storm spotters was tornadoes and thunderstorms. These severe weather events still make up the majority of storm spotter activations. Today storm spotters are also activated for a wide range of severe weather events such as hurricanes, winter weather, floods, dust storms, and nor'easters. Spotters are also activated for geologic issues such as coastal erosion and volcanic ash fall.

Over the last 30 years there have also been monumental advancements in technology that have benefited storm spotting. Within the Amateur Radio community the growth in repeaters on the 2 meter and 70 centimeter bands have made running SKYWARN nets possible almost anywhere in the country. And new modes such as *Echolink, D-Star,* and *APRS* have helped to fill in communication gaps and generally make storm spotter networks more robust.

Outside of Amateur Radio there have been many advancements that have helped volunteer storm spotters become more effective. Cell phones, once a rarity, are commonplace. They can provide voice as well as data communication in many areas. Computers have become an essential tool for the storm spotter. Combined with modern weather software packages they give the storm spotter or net control operator access to data not available 30 years ago. Dissemination of weather information has increased exponentially. Today weather information can arrive via television, radio, satellite, NOAA Weather Radio, text message, e-mail, and Web sites. Instant messenger programs make it possible to network spotters, net control stations, emergency management, meteorologists, and media seamlessly.

With the advancement in technologies also come advances in what we have learned from severe weather. Regardless of your role in the severe weather warning process - spotter, public safety official, NWS, news media - with each severe weather event we learn from it and learn how to respond better in the future.

The ability to forecast severe weather has also evolved greatly over the last 30 years. Advances in radar, communications, networking, and research have made forecasting more accurate. While forecasting is not a 100% accurate science, you can look at the records of fatalities and injuries related to severe weather and see that progress is being made.

For the Amateur Radio operator who is a storm spotter the amount of available technology, resources, meteorological information, training, and the impact from severe weather can all seem daunting. In this book we will look at these things and see how Amateur Radio operators can apply the right combination of tools to be safe during severe weather, respond to it, and assist in its aftermath.

OUR PURPOSE, AUDIENCE, AND ASSUMPTIONS

Before we get into the heart of the material we need to cover the purpose and intent of this book and who it is written for. We also have to go over certain assumptions that will be made throughout this book. And finally we need to make some clear distinctions about storm spotters.

The purpose of this book is *not* to compete with the NWS's official SKYWARN spotter training. Storm spotters, whether they are Amateur Radio operators, public safety officials, or just concerned citizens, must be trained by meteorologists. We must remember that the purpose of a storm spotter is to relay ground truth information to those that issue our nation's warnings and forecasts, the National Weather Service. This is why the NWS developed SKYWARN training and why NWS meteorologists conduct SKYWARN training. Granted, there are other training programs out there designed for storm spotters, but all will generally agree that the basis for training comes with the basic SKYWARN course. And many that go through this course are Amateur Radio operators. This is not by accident. Amateur Radio and storm spotting are a great match. Amateur Radio operators bring to storm spotting great resources; an established communications system that can function in an emergency, a pool of volunteers willing to be trained, a history of public service, technologies that no other group has, and all paid for by the individual amateur.

The purpose of this book is to be a resource for the Amateur Radio operator who volunteers as a trained storm spotter. It presents information on resources, training, equipment, safety, storm spotter activation procedures, reportable weather criteria, developing a local storm spotter manual, and the experiences of storm spotters from around the country. It also provides some meteorological information about severe weather such as hurricanes, tornadoes, hail, floods, damaging wind, and winter weather.

The primary audience in mind is the Amateur Radio community. It is not just for experienced storm spotters but also for those new to this aspect of the hobby or interested in getting into storm spotting. Non-amateurs may also find some material useful. While they are not the intended audience if the material presented helps them as a storm spotter or perhaps generates interest in Amateur Radio than that would be welcome.

There are certain assumptions that we must make throughout this book: definition and purpose of a storm spotter, storm spotter affiliation, function of storm spotter reports, the unique role that Amateur Radio plays, and fundamental training of all storm spotters. There is no single, set definition of what a storm spotter is. We can define a storm spotter, though, in these terms: a storm spotter is a volunteer who visually monitors specific weather conditions and their progression and relays that information to a local weather office. In the United States these volunteers are trained by the National Weather Service through the SKYWARN weather spotter program. The NWS defines the responsibility of the storm spotter as "identify and describe severe local storms".[8] The purpose of having this pool of trained volunteers is to provide ground truth reports which along with other partners, such as media, storm chasers, public safety officials, and emergency management, make up a reporting network. Information from this network is used, along with information from radar and other sources, in issuing severe

Storm Spotting & Amateur Radio

20 Questions for a Storm Spotter

Q1. How long have you been a storm spotter?
A. 10 years.

Q2. What types of severe weather are you activated for?
A. Any weather that has the potential of becoming severe that could directly affect property and life.

Q3. What type of training have you had?
A. Training is offered through the NWS to any persons that are interested in becoming a spotter for the NWS. Check with your local amateur radio group or NWS office to find out more about a class near you.

Q4. Approximately how many storm spotters are in your area?
A. 10 to 15.

Q5. How does information get from the spotter to the NWS?
A. The most common way is through the use of amateur radio. Amateur radio plays a vital part in passing along severe reports directly to the NWS.

Q6. Does the local storm spotter group participate in any drills or exercises?
A. Yes. As in many cases, drills keep spotters in practice, and prepared for the real thing.

Q7. How does your local SKYWARN group include non-ham spotters?
A. While Amateur Radio has a big part in SKYWARN, anyone can participate, even if they don't have a ham license. Reports can be phoned into the NWS as well.

Q8. How often do you have local SKYWARN training?
A. Training classes are generally held from December through February, just before the storm season begins.

Q9. Describe the relationship between your SKYWARN group and local emergency management.
A. We assist local agencies in attempting to provide an early warning of approaching sever weather. It's all about saving lives!

Q10. What Amateur Radio modes and frequencies do you use for storm spotting?
A. Most areas have local SKYWARN nets that have a net control at the NWS. In some rural areas, however, a spotter may not be able to communicate with the net control. In those cases, the NWS is also monitoring an HF radio close by.

Q11. As a volunteer storm spotter what are your primary safety concerns?
A. The main concern for any spotter is the effort to save lives. In order to do that, communication is necessary to relay reports of sever weather that could have an impact on human life. Amateur radio plays a vital role in the community, providing an early warning when threatening weather approaches, thus allowing people a better chance getting to safety.

Q12. In your area are the evacuation centers, Red Cross, EOCs, etc equipped with Amateur Radio?
A. It is an ongoing effort to have a station where ever possible to assist in emergency communications. Most amateur radio operators can set up an emergency station very fast, if one is not already in place.

Q13. Describe how your local SKYWARN net is conducted.
A. All reports are communicated directly to the net control who then passes along the report to the NWS. Scheduled nets are usually held weekly, to keep in practice for an actual net.

Q14. Does your local SKYWARN group conduct or participate in any training that is not SKYWARN?
A. Yes. ARES / RACES also play a big part in emergency communications. Contact your local amateur radio group to learn more.

Q15. What other Amateur Radio activities are you involved in?
A. Amateur Radio operators not only participate in emergency communications, they participate in other activities as well. Field Day is an excellent opportunity to witness and even get involved in the various operations of Amateur Radio.

Q16. What got you interested in storm spotting?
A. Participation in the local Amateur Radio club.

Q17. Do you have a go-kit? What's in it?
A. Yes. A portable ham radio, batteries, bottled water, first-aid kit, flash light, etc. A go-kit is a good idea to have in place in emergencies.

Q18. What, in your opinion, is the most valuable tool a storm spotter can take with them when spotting?
A. Radar. Utilizing radar allows spotters to know where a storm is, where it is headed, and what parts of the storm are severe.

Q19. How can Amateur Radio storm spotters improve?
A. It is the goal of every amateur radio spotter to keep room open for improvement. Any way a ham can better serve his or her community is certainly the desire for every amateur radio spotter.

Q20. When you are activated how do you participate in SKYWARN? (ie mobile spotter, net control, relay, etc)
A. It is important that a spotter is able to rotate through each area when an activation occurs. All play a vital role in saving lives including participating as net control, relaying information, or even tracking a storm in the mobile. It is important to also have the proper training and participate in drills to have proper knowledge of how and what to do should an actual SKYWARN activation occur.

Answered by Matthew Utley, KD5HAO, Wynne, Arkansas

Storm Spotting & Amateur Radio

20 Questions for a Storm Chaser

Q1. How long have you been a storm chaser?
A. About 16 years.

Q2. What is your educational / training background?
A. Most of it is self taught, but I have taken many classes in Earth Science, meteorology, math, chemistry and geography.

Q3. How would you define the difference between storm chasers and storm spotters?
A. Spotters wait for the storm to come to them and then report their findings to the NWS by some means. Chasers will track a storm down, tape it or get photos of it. Some do it for research and others for the adrenaline rush or personal gain.

Q4. Describe the relationship between storm chasers and storm spotters?
A. Most chasers started out as spotters and evolved into chasers for different reasons.

Q5. While chasing, do you monitor Amateur Radio communications?
A. Always.

Q6. What equipment do you take chasing?
A. Laptop with Internet, water and snacks, cellphone, go-kit, camera, batteries, ham and CB radios, NOAA radio, pens, paper and basic office supplies. We used to pull over and use a payphone to call in reports and had paper forecast charts that were printed out the night before to use. It sure has come a long way.

Q7. What got you interested in chasing?
A. The science of how a tornado is formed. It's pure fascination.

Q8. When chasing, do you coordinate with local public safety and emergency management officials?
A. Yes.

Q9. How do you relay information to NWS?
A. Via ham radio or phone.

Q10. As a chaser what are your primary safety concerns?
A. The other people out there when we are chasing. They tend to stop suddenly and drive towards what they are staring at.

Q11. What types of severe weather do you chase?
A. Thunderstorms and tornadoes.

Q12. In your opinion, do you think storm chasers can help in training storm spotters?
A. Yes, we have real-time experience with these storms that most spotters may never see.

Q13. How many are in a chase team and what are their roles?
A. The number varies. My team has three. Driver, navigator / radar man, and cameraman. Most chasers, however, meet up with many others in the field so the number may be as high as eight or more in a vehicle doing different stuff.

Q14. What geographic areas do you storm chase in?
A. I've chased from coast to coast and border to border.

Q15. When not chasing what do you do? (ie meteorologist, academic, etc)
A. I have a full time job at the local newspaper.

Q16. What kind of relationship is there between storm chasers and media?
A. That's tough. We all have certain media outlets we use to sell video or pictures for the local news. We very rarely go outside that circle.

Q17. How do you determine when to go on a chase?
A. Well, if the models show a decent area for storm development, the decision is made based on models and gut instinct. Usually you know a couple of days in advance.

Q18. Is there specialized training for storm chasers?
A. Not really. Most are self taught.

Q19. How realistic are TV shows about storm chasing?
A. Like the show Storm Chasers on Discovery Channel? I would say it's a fifty-fifty mix. They edit out the days you bust because it doesn't make good watching. There are days on end where you drive and drive hundreds of miles to get to an area that looked great for development a couple of hours earlier only to sit for hours and do nothing and have the storm die out before you see anything. They never show that . . .

Q20. In what ways can the field of storm chasing improve?
A. That's hard to answer. There are so many factors taken into account for chasing on any given day that to narrow it down is nearly impossible. It's not a perfect science. I guess the image itself of a storm chaser could be improved. There are so many people who have seen that famous movie and watched all those shows and then saw that guy get so close to the tornadoes that he could touch it. Because of this, they are perceived as reckless, dangerous, law-breaking people. Others who don't have experience tag along behind our groups and think they can do it too. They don't realize how dangerous it truly is until it's too late and someone doesn't come home.

Answered by Benjamin Rock, WX9TOR, Woodstock, Illinois

weather watches, warnings, and advisories. Spotters trained and certified by the NWS are considered volunteers to the Federal Government. A non-paid volunteer observer may be considered an "employee" under the Federal Employees Compensation Act. Final determination rests with the Department of Labor's Office of Workers Compensation Programs. Any spotter injured while providing observational duties should notify their local NWS office. The local NWS office and their regional office should work with the Office of Workers Compensation Programs for resolution.[9]

It is absolutely critical that we understand an important concept about storm spotting. Storm spotters are *not* storm chasers, and likewise, storm chasers are not storm spotters (see the sidebars on pages 10 and 11). These are two different activities that are often, and mistakenly, considered to be the same thing. There are very critical differences between these two activities. First is the difference in what weather conditions each one is looking for. Storm spotters are observing severe weather conditions in their area. The weather conditions that are reportable are set by the local NWS office. Storm chasers are actively seeking and going after severe weather which may be occurring far away from where they live. Training is another difference. Storm spotters are almost always trained by the NWS SKYWARN program. They may also receive additional training in severe weather through advanced SKYWARN or through local emergency management or other sources. Storm chasers, many of whom are meteorologists, generally have a broader, more in depth knowledge of severe weather. This is not to say that spotters are less knowledgeable, but that the difference in what each one does requires different levels of meteorological knowledge.

Storm spotters are coordinated through the local NWS office. When spotter activation is necessary the local office will let spotters know through hazardous weather outlooks, e-mail, text messaging, or via local emergency management. At the community level spotters are often coordinated through Amateur Radio clubs, emergency coordinators, SKYWARN coordinators, or emergency management. Storm chasers most times operate independent of any organization although some are coordinated through a university or research center. While in the field chasers may coordinate and share information with other chasers.

Equipment is also a major difference. Storm spotters generally use basic equipment that helps them monitor local weather conditions. This may include and amateur radio, cell phone, camera, and GPS. Storm chasers often use professional grade equipment to monitor a wide range of weather conditions. Storm spotters operate within in local area. They are trained to report on conditions unique to where they live. Storm chasers may travel hundreds or thousands of miles to chase severe weather.

The information gathered by spotters and chasers can be used for a variety of purposes. A spotter's primary responsibility, as stated earlier, is to relay important weather information back to the local NWS office. This may be done directly or via a net control station, relay station, or emergency management. Storm chasers may also report weather information to the local NWS office, but often the data they are gathering may be used for research purposes, commercial purposes, or individual interests.

There is one critical difference that every storm spotter should keep in mind. While the NWS does support storm spotters through the SKYWARN program, it does not support storm chasing. It is absolutely critical that every storm spotter know this and know the boundaries between spotting and chasing. This book is intended for storm spotters, not chasers.

Amateur Radio and SKYWARN have enjoyed a solid relationship since the inception of SKYWARN in the late 1960s. This relationship has benefited the Amateur Radio community by providing training, a regular source for public service, realization for the need to advance communication capabilities, and a solid relationship, not only between the ARRL and the NWS, but all the way down to the local Amateur Radio club and the local NWS office.[9] For the NWS this relationship has provided a large pool of volunteers that come with a communications network designed for emergency use, a vast increase in the number of storm reports, improved communication during severe weather, and another link to the local community. Simply put, without spotters the NWS would not be able to fulfill its mission of protecting life and property.

Now let's get into the heart of the material. The first issue we need to address, and will continue to address throughout this book, is safety. Safety is *always* our #1 priority.

REFERENCES

[1] D Ludlum, "Early American Tornadoes, 1586-1870," American Meteorological Society, 1970.

[2] National Weather Service, "NOAA's National Weather Service, Cooperative Observation Program": **www.nws.noaa.gov/om/coop/what-is-coop.html**, 2009

[3] C Doswell, A Moller, H Brooks, "Storm Spotting and Public Awareness Since the First Tornado Forecasts of 1948", Weather and Forecasting, 14, 544-557, August 1999.

[4] H Mason, "Amateur Radio Furnished Communication During Flood," *QST*, August 1923.

[5] "Amateurs Provide Communication During Ice Storms," *QST*, March 1937.

[6] C DeSoto, "Texas Hurricane Finds Hams Ready," *QST*, November 1941.

[7] "So Carolina Tornado," *QST*, January 1939.

[8] National Weather Service, "What is SKYWARN": **www.nws.noaa.gov/skywarn**, 2009.

[9] Federal Employee's Compensation Act, 5 USC Chapter 81

Chapter 2

Safety

On May 10, 2008 a series of supercells developed over northeast Oklahoma, southeast Kansas, and south west Missouri. The storms produced hail up to softball size, lightning, damaging winds, and 11 tornadoes. The storm system was part of a larger tornado outbreak that spawned 147 tornadoes over 18 states between May 7 and May 15. Altogether the outbreak resulted in 25 deaths; 22 of the fatalities occurred in Picher, Oklahoma, and Newton, Jasper, and Barry Counties in Missouri.

One of the fatalities was a storm spotter. Tyler Casey, a volunteer firefighter from Seneca, Missouri, was killed when his vehicle was thrown by a tornado. Tyler was spotting near the intersection of Highway 43 and Iris Rd, not far from where an EF-4 tornado touched down.

The purpose of mentioning Tyler's untimely and unfortunate death while serving his community is not to open it up for critique or review. It is, instead, to drive home the point that severe weather is dangerous and can injure and kill. When we serve as storm spotters, whether at home, from a net control location, or while mobile, we take certain risks. Our first duty and a duty above any other (including submitting reports), is safety. We must keep ourselves safe and we must not do anything that would jeopardize anyone else's safety. Safety is a topic we can never go over too much. It should always be on our mind when serving as a volunteer storm spotter.

We must keep in mind that storm spotters do not have to operate mobile. Being mobile does give us the advantage of observing incoming weather from a better vantage point and the ability to leave the area if it becomes too dangerous. But overall storm spotting while mobile is far riskier than spotting from a fixed location. There are times that storm spotting while mobile is too dangerous to do no matter how experienced or well trained the spotter may be. Hurricanes, low visibility conditions such as fog or dust storms, blizzards, and spotting at night are all times when a spotter should not venture out but report from a secure location. But just because we are in a house or some other structure does not guarantee safety. Generally we are safer inside a building than in a car but no building is immune to the effects of weather. Homes, fire stations, hospitals, police stations, and even emergency operations centers have all been damaged or destroyed by severe weather. In 2005 the Knight Township Volunteer Fire Department in Vanderburgh County, Indiana was heavily damaged by a tornado.[1]

In this chapter we will look at some guidelines that can help keep a storm spotter safe. The National Weather Service, working with the American Red Cross, has developed spotter safety guidelines that are a key part of the SKYWARN training program. Beyond that there is safety advice that should be heeded from a public safety standpoint. And there are lessons learned from those that have served as spotters and even from the storm chasing community. We will look at all of these and how they apply to storm spotting. We will then look at safety recommendations that are specific to different types of severe weather. And finally we will look at items you can keep in your car or in your home that will help keep you safer.

MOBILE SAFETY

First let's look at safety

A supercell advances across Kansas.

issues for the mobile storm spotter. A mobile storm spotter, while able to gain the best vantage point for observing weather conditions, is also at the greatest risk if something goes wrong. A vehicle provides poor protection against hail, wind, and tornadoes. And getting stuck in a dangerous situation is a possibility. Storm spotters can help minimize some risks by following certain safety guidelines, taking some preparatory measures, and making sure their vehicle is properly equipped.

Buffer Zone

When spotting while mobile you should always keep a buffer zone between you and the storm. By keeping a safe distance you have more room to maneuver if the storm changes direction. Also it allows you to keep an open escape route if you need to leave the area. In the picture below the spotter sets up in the rear right flank of the storm. Since the storm is moving from the southwest to northeast the spotter is in a vantage point to observe the storm without being directly in its path. They are still able to observe reportable conditions. This location also gives them several escape routes if they need to leave the area.

Critical in understanding the buffer zone to know what else may be coming. There may be another dangerous storm behind the one being observed. Net control can help by relaying information to the storm spotter, keeping them updated on what else they can expect. Net control should relay to the mobile storm spotters what is heading their way, any associated storm reports, changes in storm speed or direction, expected time of arrival, and perhaps even a recommended safe location to move to. Net control can do a lot to help keep spotters aware and safe during severe weather.

Spotting in Pairs

At no time should a storm spotter operate alone while mobile. It is best to spot in pairs with one focused on driving and the other on spotting. It is dangerous enough to drive near severe weather, even more so if distracted by trying to storm spot at the same time. While out spotting the driver not only has to deal with adverse road conditions but also be alert to others on the road, following traffic laws, and keeping aware of escape routes. The driver's only task should be driving; doing this will allow the spotter to focus attention on observing weather conditions. The spotter should also be the one that takes care of any equipment in the car such as radios, cameras, cell phones, etc. Every effort should be made to minimize distractions to the driver. Both, or all in the vehicle, need to keep alert to potential hazards. Safety is everyone's responsibility.

Driving and Vehicles

The driver assumes a great deal of responsibility when out spotting. There are hazards on the road during severe weather that the driver must stay alert for. The driver must keep alert for emergency vehicles that will be on the road. Severe weather often calls for a greater response from public safety officials. They may be responding to a fire caused by lightning, power lines down across a road way, a submerged vehicle, or a traffic accident. Public safety responders must respond in a timely manner. When visibility is limited and it may be harder to hear an oncoming siren due to rain, wind, or noise from inside the vehicle, the driver may not be aware of an emergency vehicle as they would under normal conditions. Besides emergency vehicles the driver must be aware of other traffic on the road including pedestrians. Remember visibility may be limited and this can affect reaction time. Severe weather can also create road hazards that may not normally be encountered. Flood waters can wash away roads and bridges, ice and snow can cause roadways to be slick and hazardous, and strong winds can blow a vehicle off the road. It is also advisable to avoid gravel and dirt roads if there has been intense rain or flooding. These roads may become soft causing vehicles to get stuck. It is better to observe from paved roads and better still from four lane intersections. These will provide the greatest number of possible escape routes on good roads.

Along with being aware of what is on the road the driver must remember to observe all the rules of the road. It is too easy to get in a hurry when storm spotting. You should drive at speeds appropriate for road conditions and never in excess of the posted speed limit. And everyone in the vehicle must wear their seat belts.

The driver should also never attempt to drive through the core of the storm. Storm chasers call this punching the core. By driving through the core of the storm you run the risk of encountering deadly hail, intense rain, and if there is a tornado present you may

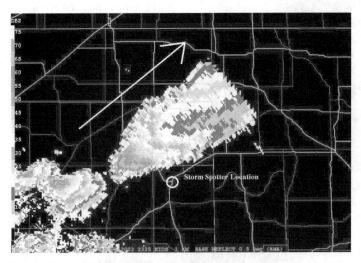

Using the radar imagery from the Greensburg, Kansas tornado (storm direction is indicated by the arrow) we can identify a good spotter location. There are several roads that lead away from the storm and can be used as an escape route.

A severe thunderstorm bears down on East Texas.

not know it until it's too late.

If you do have to stop the vehicle there are also certain safety precautions that should be taken. First is to make absolutely certain that you are out of the travel portion of the roadway. While the shoulder of the road is usually a place where a vehicle can pull off it can be a dangerous spot if visibility is limited. Try to find some place like a parking lot or rest stop area where you will be out of the way of traffic. Also avoid trees, power lines, and large signs. Strong winds can knock these down and pose a hazard to anyone nearby. Do not take cover under a highway overpass. Strong winds or tornadoes can send debris under the overpass and can result in injury or death. If conditions get too dangerous get inside a sturdy structure.

As mentioned before spotting at night is far more risky than spotting during daylight hours. Visibility is of course lower due to low light. This may make it difficult to see what is coming your way. Lightning may only give you momentary glances of the storm, but at the same time can also cause brief flash blindness. Not being able to see the storm will also make it more difficult to maneuver around the storm.

Besides driving, road hazards and conditions, and locating a safe spot and escape route, you must also make sure your vehicle is maintained properly. Not taking simple maintenance steps may pose safety hazards when out spotting. One of the most important things is making sure you can see what's coming at you when it's raining. Make sure wiper blades are good and applying something like *Rain-X* can help keep the windshield clear. Your vehicle needs to be in good mechanical condition. Make sure all fluids are filled to the appropriate level, tires properly inflated, and that all the lights work. And you need to make sure you have enough fuel. As a general rule you shouldn't head out spotting with less than half a tank of gas. Keep in mind that you may not be taking a direct route to where you are going. And if roads become flooded or blocked there may be many detours getting back. And while out spotting keep the vehicle's engine running when stopped as a precaution.

There are also things that you can keep in your vehicle to help you stay safe while out storm spotting. It should go without saying that Amateur Radio operators will have communications equipment in the vehicle with them, but don't forget extra batteries for the handheld transceiver, the cell phone and car charger, a list of important phone numbers, and a repeater directory. Also be sure to keep a basic first aid kit in the vehicle as well as a fire extinguisher. Have a blanket available in the vehicle. In winter weather it can be used to keep warm if you experience car trouble and in the event of a hail storm it can be used to provide a little additional protection. Make certain that you have a flash light and batteries in the vehicle too. And don't forget to have NOAA Weather Radio with you.

When you are storm spotting mobile you have to exercise extreme caution and be prepared for all hazards that you may face. A vehicle provides less shelter than a structure but this can be compensated for, to a certain degree, by keeping distance between you and the storm, planning an escape route, spotting in pairs, and making sure you and the vehicle are prepared for spotting. For further reading on the subject of mobile safety during severe weather see *Storm Chasing with Safety, Courtesy, and Responsibility*[2] by Charles Doswell III of the National Severe Storms Laboratory. Although written by a storm chaser it has many valuable lessons for storm spotters.

SPOTTING FROM A FIXED LOCATION

Some storm spotters will report from a fixed location such as a home, EOC, fire station, or shelter. Regardless of where they are at there are also certain safety precautions they should take during severe weather. Fixed locations offer a greater degree of shelter than a vehicle. There is a trade-off though. When in a vehicle you can leave a specific area quickly if it becomes suddenly hazardous. When in a home or other fixed location you must keep aware of what is heading your way. If you have enough time you may be able to move to a safer location. But if the weather changes suddenly you may have to stay put and take cover. The fixed location storm spotter should have a plan for dealing with each potential weather hazard. There are several safety considerations the fixed location spotter must keep in mind.

The Location

The first thing the spotter must consider is the location itself. Before severe weather strikes the spotter should do an assessment on their location to determine weaknesses and strengths. At what point will wind damage the structure? Is it in a flood plain? What is the water table like? What damage is possible from falling trees or power lines? We assess our location based on the likely threats we will encounter.

Lightning is one of the greatest dangers when operating outdoors. In the United States 200 people lose their lives to lightning strikes each year.

If I reside in the Midwest I will likely assess the impact of threats such as tornadoes, floods, and winter storms. The threats you will face depends a lot on where you are at.

Along with the structural concerns Amateur Radio operator storm spotters must also consider antennas and towers. Whether we put up a tower and antennas or string antennas up from trees and other supports we must keep in mind how severe weather will impact them. Strong winds can take down towers and antennas so you will need to make sure they don't risk hitting power lines or structures. Ice can accumulate and damage antennas and towers. Flooding can cause ground to break away. And a lightning strike can not only destroy towers and antennas but also strike the house.

Operating

Staying on the air during severe weather can be a tremendous challenge. For the storm spotter staying on the air, online, or on the phone to submit storm reports is essential. But staying operational during severe weather also requires awareness of certain safety issues. The greatest safety issue we face operating during severe weather is lightning. Lightning gives no warning that it is going to strike a specific location. But there are precautions that can be taken to minimize the damage caused by a lightning strike. Refer to the appendix for a guide on lightning protection for the Amateur Radio station. Here are a few safety tips to follow when lightning is present while active as a storm spotter from home.

- Stay off of corded telephones. Lightning can travel along telephone lines. If you need to phone in a weather report use a cell phone.
- Use a handheld radio. Even with the best lightning protection it is best not to tempt fate. Avoid using a radio that is connected to an outside antenna.
- Avoid items that can conduct electricity. Lightning can enter the house from many sources; wires, pipes, even the ground. It does not take a direct hit for lightning to get into the home.

Being inside a house is one of the best steps you can take for lightning protection. It does not mean you are in a lightning-proof place, though. Read and follow the lightning safety guidelines available from the NWS.

Operating From an EOC or Other Facility

Often during SKYWARN activation, and ARES activation for that matter, net control will be located at an Emergency Operations Center (EOC). There are some distinct advantages to this. It allows net control to stay in touch with critical players in emergency response, such as emergency management, public safety, public works, and government officials. This close contact allows important weather information to be disseminated quickly to key individuals. Operating from an EOC also allows a dedicated station to be set up for emergency operations. Net control also may operate from fire stations, hospitals, shelters, or other similar locations. The general idea for all of these is to keep net control close to important decision makers and make it possible to relay important information quickly.

For our purposes we will assume net control is operating from an established EOC. The safety guidelines for out EOC net control will be similar to those that net control stations at other locations should follow.

Operating net control at an EOC is a lot like operating

from a public safety dispatch center. It is a dedicated station. It operates during emergencies and crises. It is manned by an operator. And it has to stay on the air. To do this safely the net control operator is going to want to find out a few things. First, is that station grounded? Is there lightning protection? What kind of power back up is there - generator? batteries? What station equipment is connected to the back-up power system? Is there an evacuation plan and is it posted? Is there a first aid kit handy? Is there back-up Amateur Radio equipment? The net control operator must know the answers to these questions. Just because we're located at a station, in a secure facility, designed to stay on the air doesn't mean we are immune from the effects of severe weather. Lightning can still strike and cause injury or death and damage equipment. The EOC can be in the path of the tornado. In 2003 an EOC in central Indiana was inundated with flood water because it was located in a basement. Amateurs that volunteer as net control in EOCs and similar facilities should be familiar with potential hazards. Talk with the EOC manager or other official in charge about any risks the center faces. Talk with other amateurs that have operated as net control from that EOC and learn about their experiences. Just because it is an emergency operations center does not mean it is secure. Many EOCs have been placed in less than ideal spots as ways for jurisdictions to save money or as the result of poor planning.

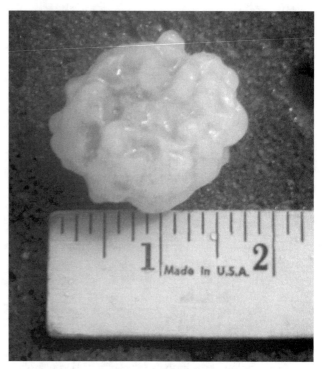

An example of potentially dangerous hail.

NWS WEATHER SAFETY INFORMATION

For decades the NWS has produced information in the form of pamphlets and flyers, videos, Web sites, and in its actual severe weather watch, warning and advisory information. Safety information is also included on NOAA Weather Radio broadcasts. Weather safety information is one of the most important NWS products.

Currently the NWS produces safety information on lightning, hurricanes, thunderstorms, floods, tornadoes, heat, and winter weather, just to name a few. Every SKYWARN storm spotter should be familiar with severe weather safety. We cannot get into the frame of thought that we know enough just because we took the SKYWARN training. Read and heed the safety guidelines from the NWS (see the sidebar on page 18). No storm spotter should *ever* compromise safety to turn in a weather report.

The National Weather Service forecast center in Charleston, South Carolina.

Storm Spotting & Amateur Radio

NWS Weather Safety Information
The following weather safety information has been provided by the NWS for this publication.

Tornado Safety
- Be on the lookout for other tornadoes that could form in the vicinity of the tornado you are watching.
- Never try to out-run a tornado in an urban or congested area, but immediately get into a sturdy structure after parking your car out of the traffic flow.
- Do not take shelter under bridges or overpasses. These structures do not offer ample protection and could *increase* the chance of injury or death.
- If you are caught outdoors, seek shelter in a basement, shelter or sturdy building. If you cannot quickly walk to a shelter, immediately get into a vehicle, buckle your seat belt and try to drive to the closest sturdy shelter. If flying debris occurs while you are driving, pull over and park. Now you have the following options as a last resort:
  o Stay in the car with the seat belt on. Put your head down below the windows, covering with your hands and a blanket if possible.
  o If you can safely get noticeably lower than the level of the roadway, exit your car, and lie in that area, covering your head with your hands.
  o Your choice should be driven by your specific circumstances.
- Flying and falling debris is the biggest hazard in a tornado. To be safe, you should get in, get down and cover up. Underground or in a safe room is best. If no underground shelter is available, get to the center of a sturdy building on the lowest level. Put as many walls between you and the tornado as possible. Stay away from windows and doors. Cover up to help minimize being injured by flying or falling debris.

Flash Flood Safety
"Turn Around, Don't Drown!"
- Do not attempt to drive or walk across a flooded road or low water crossing. You cannot be sure about the depth of the water or the condition of the roadway - it might be washed out.
- Two feet of moving water will carry away most vehicles.
- Six inches of fast-moving water can knock you off your feet.
- If your vehicle is suddenly caught in rising water, leave it immediately and get to higher ground.
- Be especially careful at night when flash floods are harder to recognize.

Lightning Safety
"When Thunder Roars, Go Indoors!"
- Remain in a hard-topped vehicle or an indoor location for at least 30 minutes after you hear the last thunder clap. If you use radio equipment, avoid contact with it or other metal inside your vehicle to minimize the impacts should lightning strike.
- If you are out on the water and skies are threatening, get back to land and find a fully enclosed building or hard-topped vehicle. Boats with cabins offer a safer, but not perfect, environment. Safety is increased further if the boat has a properly installed lightning protection system. If you are inside the cabin, stay away from metal and all electrical components.
- When reporting to the NWS, use a cordless or cell phone if available.
- Lightning victims do not carry an electrical charge, are safe to touch, and need urgent medical attention. If a person has stopped breathing, call 9-1-1 and begin CPR if the victim is not breathing.

Downburst Wind Safety
- Keep a firm grip on your vehicle's steering wheel to maintain control. Downbursts can occur suddenly with an abrupt change in wind speed and direction.
- If you can do so safely, point your vehicle into the wind to minimize the risk of the vehicle being blown over.
- Be prepared for sudden reductions of visibility due to blowing dust or heavy rain associated with downbursts.
- Point spotters observing from a substantial building should move away from windows as the downburst approaches.

Hail Safety
- Substantial structures, like a garage, offer the best protection from hail.
- Spotters in vehicles should avoid those parts of the storm where large hail is occurring.
- Hard-top vehicles offer good protection from hail up to about golf ball size. Larger hail stones will damage windshields.

Used with permission of the National Weather Service

Flooding can easily catch you unaware.

SPOTTER SAFETY OFFICER

Training and severe weather awareness can do a lot to help keep us safe during SKYWARN activation. However, when we're out storm spotting or operating at the net control station site we can easily get so caught up in the many tasks that have to be done that we may overlook safety issues. We can borrow an idea from industry and public safety, the safety officer, or for our purposes the Spotter Safety Officer (SSO).

The SSO is appointed by either the local SKYWARN coordinator or net control station. Their duty is to keep constant tabs on storm spotter positions relevant to the storm. If conditions dictate the SSO will advise the NCS to have spotters move to a new location. To do this successfully the SSO must carefully monitor storm location using radar and information from the NWS and local new media, know where spotters are located, pay careful attention to spotter reports of hazardous weather in the area, and plot this data on a map. Because safety is at stake the SSO must be knowledgeable about severe weather, storm structure, and storm characteristics. The individual selected for SSO must be someone that spotters and net control trust to look after their safety. It is not a position to be taken lightly.

The success of the SSO is dependent on the storm spotters. Spotters must go to their assigned locations and only move at the direction of net control, when safety dictates, or when activation is ended.

A final duty of the SSO is to advise the net control station or local SKYWARN coordinator on safety concerns. If an individual storm spotter is acting in an unsafe manner the SSO should advise NCS or the SKYWARN coordinator, who can then take action.

In the same way Amateur Radio polices itself, so too should those involved in SKYWARN activities. There is no room for unsafe behavior.

REFERENCES

[1] M Mitchell, "Indiana Tornado", *9-1-1 Magazine*, May 2006.
[2] Charles A Doswell III, "Storm Chasing with Safety, Courtesy and Responsibility":
www.cimms.ou.edu/~doswell/Chasing2.html

Chapter 3

Equipment & Resources

When the local SKYWARN group is activated during severe weather many individuals come together for a common purpose, to observe and report severe weather conditions to the National Weather Service and other served agencies. To do this job, though, requires the right tools. Whether you are a home-based spotter, a net control station, or a mobile storm spotter, having the right equipment can make you more effective.

In this chapter we will look at the wide range of gear the storm spotter may use. We will look at mobile and base radios, antennas, GPS, lights, 'go-kits', weather equipment, and other equipment that can be useful. We will go beyond the radio to see what other items may be helpful when storm spotting. We will also look at software for the storm spotter. And we will see how modes such as *D-Star* and *APRS* can be useful to the storm spotter.

So let's get started with the most important items, radios and other communication devices.

RADIOS

The radio is, for the amateur that is storm spotting, probably the most essential piece of equipment. Whether he is acting as net control, at home, or in the car the radio is the fail-safe link to call in reports of severe weather.

There are many factors that need to be considered when we decide which radio will work best. So let's look at the considerations for home-based spotters, net control stations, and mobile spotters.

Which radio to use of course depends on what role you are taking as a spotter and what frequencies and modes you will need to use. Most local SKYWARN nets are conducted on local VHF or UHF repeaters.

A mobile VHF / UHF radio is a good choice whether you are a mobile spotter, net control, or a home-based spotter. Current mobile radios come in a wide range of options; single band, dual band with single or dual receive, and radios that combine HF / VHF / UHF in one package. Most operate on 12V, so are easy to connect to a power supply or battery. Handheld radios, while handy, are not the best for storm spotting. They are dependent on battery life, have a lower operating power, the stock antenna is less efficient, and the audio output of the speaker may not be high enough. They can be useful to the spotter though. Having one in the car can give the mobile spotter an extra receiver that can be tuned to the local NOAA weather channel, as a back-up radio, or to monitor local public safety (if legal to do so in your area). Net control operators may use them to monitor traffic if they have to step away from the radio.

There are times though that an HF radio will be needed for storm spotting activities. Local nets are used to relay information to National Weather Service offices. This can

SKYWARN net control station located at a volunteer fire department. Left to right, Caleb McCormick, KE5EOP, monitoring the NWS website; Larry Brown, K5LMB, handling check-ins on the SKYWARN 2m net. (Mike Corey, W5MPC,

Storm Spotting & Amateur Radio

NOAA WEATHER RADIO

The roots of NOAA's Weather Radio service (NWR) goes back to 1906, when the US Weather Bureau experimented with radiotelegraphy to speed notice of weather conditions. From the 1950s onward it was designed to broadcast weather information to the general public, aviation, and marine communities. It included forecasts, daily observations, watches, and warnings. In the 1990s the mission of NWR was expanded to include broadcasts on all hazards. Today NWR has 1000 transmitters, covering all 50 states, adjacent coastal waters, Puerto Rico, the US Virgin Islands, and the US Pacific Territories. NWR goes beyond weather information and can be used to alert the public to natural as well as man-made hazards, AMBER alerts, and 911 outages. If you tune to the NWR frequency, though, most days you will get a steady stream of weather-related information.

Today we have a wide variety of sources to receive weather information, so why is NWR still needed? A January 1975 White House policy statement that remains in effect designated NOAA Weather Radio as the sole government-operated radio system to provide direct warnings into private homes for both natural disasters and nuclear attack. NWR also remains the primary method for activating our nation's Emergency Alert System (EAS) in which emergency messages from the government are relayed to the American public through a variety of radio and television broadcast means.

Even though watches and warnings can be received from the NWS through the Internet, text messaging services, and e-mail, NWR still provides a valuable service. Twenty-four hours a day, seven days a week, you can receive constant, reliable weather information. And when a watch or warning is issued NWR has about a two minute lead time over other

Mobile set-up for voice and APRS. (Mike Corey, W5M-PC, photo)

be done by telephone, Internet, or by radio. The local NWS office may have an Amateur Radio station set up to run an HF SKYWARN net to collect reports from local nets in their area. If you are a net control station, relay station, or a home-based station you may need HF capability to send reports in to the NWS office. In some cases mobile storm spotters may need to utilize HF to relay reports to net control.

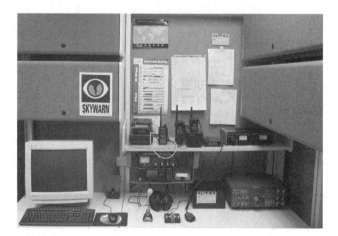

KL7FWX, SKYWARN Amateur Radio station at NWS Fairbanks, Alaska. (National Weather Service photo)

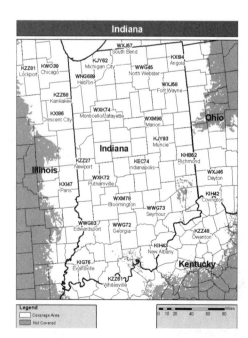

NOAA Weather Radio coverage map for Indiana. (National Weather Service)

21 Equipment & Resources

Storm Spotting & Amateur Radio

Table 3.1: The seven frequencies used by NWR.

media. This kind of heads-up can be critical when severe weather threatens. Similar to a smoke detector, an NWR can wake you in the middle of the night to alert you to a dangerous situation. Another value to NWR is that NWR coverage is almost 100% throughout the United States and its territories.

Reception of NWR does require a special receiver. There are commercially available NWR receivers that can tune to any one of NWR's seven frequencies (see **Table 3.1**). Many of these have battery back-up and Specific Area Message Encoding or SAME. SAME is a digital tone sent over normal audio using AFSK and carries an alert message for a specific area.

SAME areas are generally organized by county, parish, city, or maritime area. So if you are setting up a NWR and you live in Oxford, Mississippi you will set the frequency to 162.55MHz and program the SAME code 028071 (Lafayette County). To find your local NWR frequency and SAME code you can go to the NWR Web site.[1]

For the best performing NWR receivers, NWS suggests that you look at devices that have been certified to Public Alert™ standards. These radios meet certain technical standards and come with many features such as SAME, a battery back-up, both audio and visual alarms, selective programming for the types of hazards you want to be warned about, and the ability to activate external alarm devices for people with disabilities.

Reception of NWR is not limited to weather radios, however. Many single and multi-band VHF / UHF transceivers are capable of receiving NWR. Several have NWR as a one-touch feature with all NWR frequencies stored in a separate dedicated memory bank. While these radios may not be able to be programmed with the SAME codes, many are capable of picking up the 1050 Hz tone that is generated when there is an emergency alert. The only difference is that this is not an area-specific tone.

For the storm spotter NWR is a valuable tool. The net control station should be monitoring NWR for up to date weather conditions, watches, and warnings. Home based spotters and mobile spotters should also be able to monitor NWR. A handheld VHF / UHF radio tuned to NWR is a handy tool to have when mobile spotting since the 'rubber ducky' antenna will be able to pick up the nearest NWR transmitter in most areas.

ANTENNAS

The antennas we use for storm spotting are not too different from those normally used in Amateur Radio. There are certain considerations that must be made by the storm spotter when selecting antennas, though.

For the mobile storm spotter the two biggest factors will be frequency and gain. Most mobile spotters will likely use VHF / UHF repeaters to radio in weather reports and do so from their local area. Most mobile VHF / UHF antenna set-ups will be able to do this without a problem. But if your local repeaters coverage is limited you may want to look at a higher gain antenna.

For the home-based spotter or net control station frequency will of course be an issue, but even more of an issue is the survivability of antenna systems. Unlike the mobile storm spotter the fixed-position spotter cannot easily move out of the way of approaching severe weather; antenna systems can quickly become scrap metal when hit by damaging winds or loaded down with ice. For the home-based spotter emergency antenna systems are going to be essential. For VHF / UHF operation a field deployable antenna system can keep them on the air when antennas come down. Home-based stations can also make use of mobile antennas adapted for home use. A mobile antenna magnetically mounted to a cookie sheet can be used to stay on the air. For emergency HF operation the home-based spotter should keep wire antennas on hand for the bands they will need. Wire antennas can be stored in little space, quickly deployed, and can be cut to a specific frequency.

APRS

APRS - the Amateur Packet / Position Reporting System - is "a two-way tactical real-time digital communications system between all assets in a network sharing information about everything going on in the local area."[2] It provides amateurs with the ability to share a wide range of data not only at a local level, but by use of Internet gateways at the global level as well. *APRS* has several important features for storm spotting. Of primary interest to the storm spotter are position information, weather data, and messaging.

Probably the most widely known feature of *APRS* is the ability to transmit position reports and see these reports on a map. For storm spotters, whether mobile or home based, net

---

Break-down of a SAME Message

● **Preamble** - Coded message that contains ASCII code, originator code, event code, county-location code, alert duration, date and time, eight character ID. Up to 32 locations may be sent with one message.

● **Attention Signal** - Audible 1050 Hz alert tone. Many VHF / UHF radios can be set to pick up on this tone. This signal must last at least eight seconds.

● **Message** - Audio or video message.

● **End of Message**

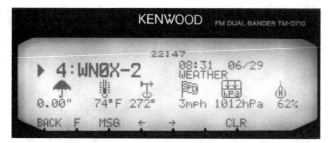

Weather information received via APRS. (Mike Corey, W5MPC, photo)

control stations, NWS offices, and emergency management this feature is invaluable during severe weather. By logging into one of the many online sites that provide APRS information such as aprs.fi it can quickly be seen who is currently on, from where, at what type of station. Net control, emergency management, and NWS can see if there are mobile spotters out and if need be reposition them or determine quickly if they are in harm's way of approaching severe weather.

Weather data can also be distributed via the *APRS* network. Using a program such as *UI-View* weather data from local Amateur Radio *APRS* stations, CWOP (Citizen Weather Observation Program) stations, and NWS watches and warnings can be seen. Mobile stations can see data being transmitted from weather stations and receive information about weather watches, warnings, and advisories.

APRS also allows users to transmit short text messages to others. They can be sent from one *APRS* user to another, an *APRS* user to an e-mail address, or even to a cell phone as an SMS text message. The Amateur Radio Universal Text Messaging / Contact Initiative is an effort underway to greatly expand the ability to send messages to users over a wide range of systems. While text messaging via *APRS* has been around for over 15 years, there are still a lot of other systems that can be integrated into a single messaging system. This is a key feature of *APRS* that spotters should be utilizing and helping to advance and develop.

While *APRS* offers so much more than position, weather, and messaging these are some critical components for storm spotters. If you have not tried *APRS* yet give it a try! For more information on *APRS* and its capabilities refer to the *ARRL's GPS and Amateur Radio* [3] and *ARRL's VHF Digital Handbook*. [4]

ECHOLINK

EchoLink has been a valuable tool to the SKYWARN community. We won't go over all the details on what *EchoLink* is here, but instead will focus on how it can be utilized by Amateur Radio storm spotters. If you are not familiar with *EchoLink* or other VoIP communications for Amateur Radio, refer to the *EchoLink* Web site[5] or *VoIP: Internet Linking for Radio Amateurs* by Jonathan Taylor, K1RFD.[6]

EchoLink and other VoIP modes such are *IRLP* (Internet Radio Linking Project) allow us to combine RF communications and the communications capability of the Internet to form a more robust network. By utilizing *EchoLink* a SKYWARN net control station can keep in touch with all local SKYWARN nets in a particular county warning area. And areas that do not have repeater coverage can still stay in touch by using *EchoLink* in an Internet-only mode connecting directly with the SKYWARN net control station at the NWS office. Utilizing *EchoLink* in this way can help fill in any communications gaps and generally make severe weather communications more effective.

Let's take a look at how *EchoLink* is used in a real situation. At the Tampa, Florida NWS office Amateur Radio is a critical part of severe weather operations. SKYWARN nets, activated during severe weather outbreaks, can be conducted at the local level using local repeaters or at a regional level using the NI4CE repeater system which has an *EchoLink* node. During regional weather events such as hurricanes the *EchoLink* node can help provide additional communications capabilities by filling in repeater coverage gaps. As long as a station can access the Internet they can still communicate with net control. It also provides a way for others outside the range of local repeaters to still access net control. For example, let's say the WX4NHC at the National Hurricane Center needs to get in touch with SKYWARN net control at Tampa WFO. Utilizing the *EchoLink* connection they can do so with only an Internet connection even though the net control station in Tampa may have no Internet connection and is operating through the repeater.

An added benefit to having *EchoLink* is during the post severe weather phase. *EchoLink* can be used as a means to conduct a regional net. Stations checking in can share reports of events that occurred in their area, problems encountered and possible solutions, and make reports to district and state emergency coordinators. In other words, it can be a conference call tool, post-event.

A great example of this is the NWS Ruskin SKYWARN Practice and Outreach Net. This net meets each Tuesday evening at 9.00pm local time on the NI4CE repeater. The net control station operates from the NWS office. The purpose of the net is "to provide limited SKYWARN™ training, make announcements of interest to SKYWARN™ personnel, provide local weather information (especially if severe weather is expected in the next few days), and to provide net participants the opportunity to interact with the staff of their local NWS weather forecast office." By combining the repeater system and the *EchoLink* node a much wider audience can be reached.

D-STAR

Over the last couple of decades Amateur Radio has seen several new communication modes develop; *APRS, PSK31, WinLink*, etc. For the Amateur Radio operator who is a storm spotter these new modes can be useful tools. We can put them to use to make our ability to communicate during severe weather more robust.

One of these new modes, *D-Star*, has been used for severe weather and hurricane response in the Southeast

Storm Spotting & Amateur Radio

United States for several years. In the sidebar John Davis, WB4QDX, District Emergency Coordinator in the Georgia Section, explains how *D-Star* works with severe weather and the Southeast D-Star Weather Net.

SKYPE™

Skype™ is another valuable VoIP tool that can help storm spotters fill in any communication gaps. *Skype* is a free downloadable program that allows users to connect to others with voice, video, and text communication. It features conference calling between multiple users. It is free to use the program, although certain features such as the ability to make and receive phone calls, a unique phone number, SMS messaging, and voice mail require a paid subscription.

Skype does require an Internet connection. This connection can come from a standard home-based broadband connection or by using a mobile Internet connection. Voice and video require a very robust connection to the Internet, so dial-up would not be recommended for this.

There are several applications for *Skype* in storm spotting. Net control stations may link together using *Skype* to coordinate spotter response. Mobile spotters may use the *Skype* connection and a web cam to relay real time video to net control, emergency management, or the NWS. The conference call feature can be used for a virtual training session between storm spotters, emergency management, and NWS. And since no license is required to use *Skype* it can be used to bridge the gap between Amateur Radio spotters and non-Amateur Radio spotters.

GPS

In the last 10 years or so GPS has come a long way. Originally designed with the military in mind it is now, for many, a part of everyday life. For the storm spotter GPS can be a valuable tool, and when combined with other technolo-

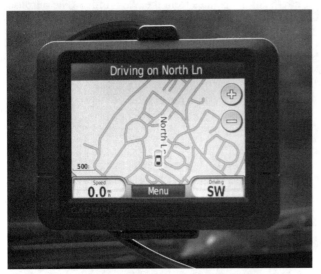

Typical mobile GPS navigation unit. (Mike Corey, W5MPC, photo)

D-STAR AND SEVERE WEATHER

Please briefly describe what *D-Star* is.
D-Star, or Digital Smart Technology for Amateur Radio, is a open protocol created by the Japan Amateur Radio League for transmission of digital voice and data. The D-Star protocol allows digital voice and simultaneous low-speed data on the 2 meter, 70 centimeter and 23 centimeter bands. In addition, it defines a high-speed data standard (128 kbps) in the 23 centimeter (1.2 GHz) band.

How many D-Star repeaters are there? How can I find out if there is one near me?
As of March 1, 2009, there are over 7300 registered D-Star users around the world. There are 385 registered D-Star systems. Each D-Star system may consist of up to four separate repeaters operated by one repeater controller, so the number is well over 1000 individual repeaters operating in the three amateur bands (2m, 70cm and 23cm).

*Two Web sites are useful for locating D-Star repeaters near you. They are **www.dstarusers.org** and **www.dstarinfo.com** The **www.dstaruser.org** Web site has a 'Last Heard' page that lists users who have accessed D-Star systems in the last 24 hours.*

What equipment do I need to get started on D-Star?
Currently, Icom is the first Amateur Radio equipment manufacturer to sell D-Star compatible radios. They are available in both several handheld and mobile models. In addition, the 'DV Dongle' is available to connect to your PC's USB port and uses the sound card to access D-Star repeaters around the world similar to EchoLink.

Are SKYWARN or any other severe weather nets run on D-Star?
There are several nets now operating on D-Star. One specifically for severe weather is the Southeastern D-Star Weather Net. It meets each Sunday night at 9.00pm ET as a training and practice net and will operate during times of severe weather in the Southeastern US, an area subject to tropical weather, hurricanes, tornadoes and severe thunderstorms. Using D-Star reflectors, which are like conference bridges, multiple repeaters and DV Dongle users can connect together for a net covering a wide area with consistent voice quality across the net and without the beeps and varying levels of traditional linked systems. An advantage of D-Star for severe weather reporting is that repeaters and

areas can be linked together quickly in whatever configuration is needed.

Are there any differences in how a net is run on a D-Star repeater over a conventional repeater?
There doesn't have to be, but D-Star does bring some unique capabilities which are advantageous for nets. Since the user's callsign is sent with each transmission, each user is identified without giving the callsign verbally. The Southeastern D-Star Weather Net uses this capability for quick check-ins using the 'Quick Key' format. When the Net Control Station calls for check-ins, users can just key their radio for one second and their callsign is visible to the net control. This allows a list of users to check-in in a shorter period of time. It can also be used for breaking stations by just keying between stations. The other stations see the new callsign displayed on their radio and can recognize the breaking station.

Also, the simultaneous low-speed data capabilities of D-Star allow 'packet speed' data to be sent with voice transmissions. It also allows for separate data to be sent using the same repeater. Comparable FM systems may require a separate radio and TNC from the voice radio along with separate repeaters or digipeaters to handle voice and data.

Are any of the National Weather Service offices on D-Star?
I don't have information on NWS office around the country, however, our plan for Georgia includes providing one in the NWS office in Peachtree City and other offices serving Georgia. Peachtree City has a well equipped and active Amateur Radio station right with the meteorologists and providing state-wide capabilities for receiving incoming reports will be valuable. Also, with the D-Star network already in place in neighboring Alabama and Florida, reports of approaching severe weather across state lines will be possible.

Please describe the severe weather net procedures on the D-Star system.
The net is still very new and procedures are being developed as we go. The primary purpose of the net is to connect affected areas with resources outside of the area. D-Star's versatile linking through reflectors makes connecting repeaters to form ad hoc networks a quick and easy process. D-Star doesn't replace other Amateur Radio resources, but is another tool in the bag bringing new capabilities to support emergency events.

Since D-Star is a network is it more suited for larger events such as hurricanes instead of smaller isolated events such as tornadoes?
Actually, the voice and data capabilities of D-Star make it equally suited for events over large areas and within a smaller, defined area such as tornado or other local event. In its simplest form D-Star is a local repeater as any, but with the clarity of digital voice. D-Star's low and high speed data capabilities over a single, or Gateway equipped, repeater can provide digital messaging to supplement voice communications or Internet connectivity within an area where traditional infrastructure may be overloaded or unavailable. During Gwinnett County, Georgia's 2008 SET, D-Star was used to relay photos from field locations directly to the EOC via the Internet.

Can D-Star support tracking and messaging capabilities like APRS?
Yes. All D-Star radios are capable of sending GPS location information or messaging built-in. The position information is transmitted in a format called D-PRS. D-Star repeaters gate the information to the Internet and to sites such as j-FindU. Unlike FM systems requiring a radio for voice and a separate radio and TNC for data or APRS, D-STAR transmits simultaneous voice and data in a single radio. With free messaging applications like D-RATS, advanced messaging, mapping data transmission capabilities are possible over D-Star.

What advantages are there in using D-Star over a conventional repeater for severe weather?
Both provide valuable services for severe weather reporting, but D-Star's linking, data transmission and position reporting expands reporting capabilities. The clarity of digital transmissions can also improve accuracy of reporting.

What disadvantages are there?
D-Star is still a relatively new technology and most amateurs still have analog radios. D-Star equipped amateurs and the availability of repeaters is growing at a very rapid pace. The cost of D-Star radios is still higher than their analog counterparts.

John Davis, WB4QDX,

gies even more so. Of course the main purpose of GPS is to answer the question "Where am I?" They also can provide accurate time, elevation, speed, heading, and tracking capability. With today's GPS units there are a wide range of uses besides location.

Today there are all sorts of GPS units available to meet a wide range of needs. GPS for the vehicle may be portable or mounted in dash. There are GPS units designed for hunting, fishing, and outdoor sports, marine use, aviation, and ones designed for motorcycles. There are also GPS units that can be put on a key fob, plugged in to the USB drive on a laptop, and several cell phones have GPS capability. For the sake of storm spotting, though, we will look at a couple of types of GPS units and how GPS can be combined with other technologies useful to the storm spotter.

Probably one of the most popular GPS devices are the navigation units sold for use in automobiles. They can be mounted within easy reach of the driver and provide a wealth of travel information; current location / speed / heading / elevation, location of nearest police station or hospital, where to get gas or something to eat, fastest route, fuel economy information, some even remind you to stop for coffee and rest. For the storm spotter these units can offer several key features. Of course the first benefit they offer is their intended use, location. The storm spotter can use the information to relay precise location of a weather event or their position relative to the event. Location can also be saved into memory with a tag for a specific weather event, known flooding area, or good location for spotting. Many also allow pictures and video to be saved into memory. A spotter can save a good spotting location and even add a picture of the view from that location. For safety purposes a GPS can give the spotter a map of the area so they can be aware of possible escape routes. Some even offer Bluetooth capability that can connect the unit to your cell phone allowing hands-free operation.

One drawback to many GPS navigation units is a lack of compatibility with *APRS*. *APRS* requires the GPS unit to output data into NMEA format.

While GPS navigation units offer the most features for the storm spotter, basic GPS units can also be useful. Units such as the Garmin ETrex can provide the spotter accurate location information. These basic handheld units can also be easily interfaced into APRS set-ups.

Several cell phones also have GPS capability either built in or as an add-on application.

GPS can function as a stand-alone tool, but to get the full benefit we need to see GPS as a multifunction tool. For the storm spotter it can serve as navigation tool, part of the *APRS* and *D-Star* systems, location information, and a potential multimedia tool.

MAPS

While GPS is a handy tool for the storm spotter to have, it is a good idea to have a bit of redundancy when it comes to navigation. Besides the redundancy factor we should keep in mind that most navigation units in vehicles are set up for the driver to see and use. During severe weather the driver should be paying attention to driving and not be distracted. Navigation is best left to a second spotter in the passenger seat. Remember safety always comes first. A good set of maps is useful to have, whether you are a mobile storm spotter, fixed, or work as a net control station. But what kind of maps should you have? Why not just keep a road atlas handy?

There are several types of maps that a storm spotter can use. First is a local street map. Although most mobile storm spotting is done in areas outside of urban congestion it is still a good idea to keep one handy. Keep in mind storm spotters also report on flooding conditions and urban areas are more prone to flooding than non-urban areas. A local street map will allow you to mark locations that are flooded and perhaps even pencil in a flooded area. City and county street maps are also handy for identifying escape routes when spotting while mobile. Remember storm spotters serve their local community, so having good local maps is important. Local topographic maps are also useful. By using the contour lines and elevations a spotter can identify a good spot to observe from and identify areas that may be particularly hazardous during severe weather such as low-lying areas. A state highway map is handy to have in order to reference weather in relation to nearby towns, counties, highways, and other points. A road atlas can also be useful for referencing weather to geographic locations. Another handy set of maps to have is a topographic / GPS atlas designed for a state or portion of a state. DeLorme has a line of these atlases for all 50 states that provide a wealth of information. Also handy are aviation maps. These are commercially available and provide aviation data on a topographic map. They cover sections of the United States each based on a major airport such as Memphis, St Louis, or Atlanta.

Something else to keep in mind is that maps don't have to be on paper. Net control stations and fixed storm spotters can utilize mapping software. New software and updates to existing software occur too frequently to bother going over specific programs. What is worth looking at are the features commonly available that are useful to the storm spotter. Most features of use to the storm spotter are available on a variety of mapping programs, so there are several choices.

Many programs feature a drawing tool that allows the user to draw lines, points, labels, polygons, or other shapes on the map. Often these can be saved for future reference. A user can use this feature to plot tornado reports, hurricane tracks, mobile storm spotter positions, or hail reports on a map. Using the polygon feature allows plotting areas of flooding. And unlike a paper map it only takes a click or two of the mouse to erase and start again.

Another useful feature is GPS interface capability. This is especially useful for mobile spotters. It can be used to plot location and route, and eliminates time spent trying to figure out where you are on the map. Many mapping programs also feature topographic maps and can provide weather data, both useful features.

Probably the most important considerations, though, will

be your operating platform and operating system. You will need to make sure that the software you choose can operate from your device; Pocket PC, Palm, laptop or desktop. Also check and make sure it is compatible with your operating system. There is software available for just about every modern platform and system.

A particularly useful piece of software for the storm spotter that deserves special mention is *Google Earth*. [7] *Google Earth* has been around since about 2001 (originally called *Earth Viewer*). In 2005 *Google* acquired the company that made *Earth Viewer* and renamed the program *Google Earth*. As of 2009 *Google Earth* is in its fifth version. The software is capable of running on multiple platforms and systems; PC, Mac, *Linux*, and iPhone to name a few. The basic software is free and there are other versions with enhanced features available for a subscription fee. Since its inception *Google Earth* has become one of the most widely used mapping programs. *Google Earth* has several features built in of use to storm spotters and a wide range of plug ins that can be added. All plug ins for *Google Earth* are in KML format. This format, which is based on XML, is the file that contains geographic data for use in *Google Earth*. Plug-ins in KML format may be written by individuals to add custom data to *Google Earth*. Many of the KML files are available in a gallery on the *Google Earth* Web site,[7] while others are available from other sites for free or for a charge.

A great feature of *Google Earth* is that it can be integrated with other programs relatively easily. See Appendix 7 on pages 134 - 138 for a guide on how to integrate *Google Earth*, APRS, and NWS data.

CELL PHONES & INSTANT MESSAGES

Today just about everyone has a cellular phone and cellular coverage is quite widespread. Over the past several decades cellular technology has also advanced beyond the original concept of a mobile telephone. Today cellular phones are capable of text messaging, connection to the Internet, video, and a wide range of other features. Many of these features can be valuable to the storm spotter. Let's look at several of these features that the storm spotter may want to consider for his tool box.

Instant messaging is a great way to exchange short messages between individuals and groups. Many schools and businesses use instant messaging to send alerts to students and employees. This feature can also be useful for severe weather preparedness. Notice of severe weather watches

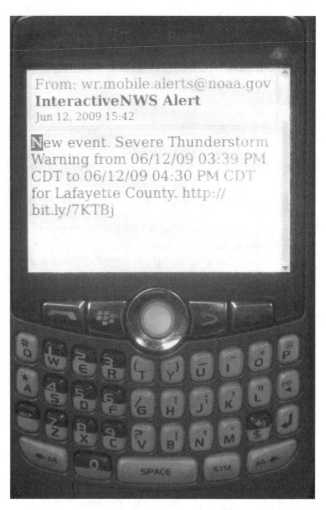

Severe thunderstorm warning from iNWS received on mobile device. (W5MPC photo)

The wind meter application for the Apple iPhone (Gery Phillips photo)

and warnings can be sent to a large number of recipients in a relatively timely manner. The NWS, The Weather Channel, AccuWeather, and several others provide weather alert instant messages. Some simply require that your phone be able to receive text messages while others, such as AccuWeather, require the phone to be Internet capable. Along with watch and warning information local SKYWARN groups may want to consider using text messaging, along with other methods such as e-mail and radio, to alert members of a SKYWARN activation. Software packages are available that allow bulk text messages to be sent.

For phones that do have Internet capability, a valuable tool, especially for the mobile spotter, is access to radar images. This can give the mobile spotter an idea of what weather might be heading his way. This is important for the mobile spotter for two reasons. First is their safety. By being able to see what is coming they can determine where to set up safely so they are not hit by an unexpected storm. Second it gives them the information they need to know where to set up to be able to get the best viewpoint. The thing that the mobile spotter must take into consideration if they are going to use this is that it adds another distraction inside the vehicle. The mobile spotter needs to keep alert to possibly rapidly changing weather conditions. They must also exercise more caution while driving due to weather conditions. For these reasons it is best to use this only when there are two spotters in the vehicle. One can keep their eyes on the road and the other acts as the spotter.

Many cellular phones on the market today come with the ability to capture still and video images. This can be quite useful to the storm spotter. By using this they can capture a video or still image of what they see and quickly forward it on to the NWS or the local emergency operations center.

There are drawbacks to cellular telephones though. Cellular networks can become overloaded during disasters, preventing voice and text communications. And as more cellular phones hit the market that are capable of voice, data, video, etc the threat of system overload increases. During a disaster emergency responders may be using cellular service to coordinate response while the general public in the affected area are making calls for help or to let loved ones know they are OK. Examples of this increase in usage can be seen during the Minneapolis bridge collapse in 2007 when cellular use was 10 times the average usage[8] and eleven times the average during a school shooting in Ottawa, Canada.[9]

Cellular technology and Amateur Radio share some commonalities. Cellular service relies on the use of designated frequencies and transceivers to handle the traffic coming in. There is a finite amount of traffic that a cellular tower can handle before it becomes overloaded. And there are other cellular sites in the vicinity that may be sharing frequency spectrum.

Despite this drawback cellular service has a distinct advantage over landline service. During severe weather landlines can be knocked down, cutting off large populations from telephone service. If a cellular site goes down another site may be reconfigured to make up for the loss.

The Amateur Radio storm spotter will typically rely on passing messages via radio. There are times, though, when relaying the message by other means may be necessary. Cellular phones can be a valuable tool for the spotter. But when cellular service goes down the spotter should know that there are still options available. The Amateur Radio storm spotter can still communicate. . . when all else fails.

LIGHTING

There are times when storm spotter activation occurs during night time hours. Storm spotting, especially mobile, at night presents even more risks to a spotter than during daytime hours. Without a doubt it is far preferable to spot from a fixed location anytime visibility is limited. However, if a spotter is mobile there are certain steps they can take to make sure they can see and be seen. And for the home-based spotter or net control station proper lighting is also critical.

First let's look at interior lighting in the mobile environment. First let's consider interior lighting in the car. Interior lighting can come from several sources; installed interior lights in the vehicle, radios, GPS navigation units, the stereo. The affect on the driver from this interior lighting can take two forms. First, it can change the driver's ability to visually adapt. Second, what is called 'veiling stimulation' where light is reflected off the wind shield. In either case there is a potential for the driver's vision to be affected negatively. When storm spotting, mobile, at night we must be aware of how lights on the inside of the vehicle can hinder our vision and ability to react to what is happening outside of the vehicle. We must remember that the driver's first responsibility is safety, not only for those inside the vehicle but also safety on the road.[10]

To help minimize visual distraction to the driver at night we must minimize interior lighting whenever possible. Mobile transceivers used for storm spotting should have their displays set to a lower light level than normal, but not to the point that they cannot be seen. The same can be said for the car stereo system. Many mobile GPS units have a night time setting that should be utilized. The vehicle's standard interior lighting should be used only when necessary. One thing that can be used to help minimize the visual affects is a flash light with a red filter. This can be handy for reading maps and adjusting radios or other equipment.

Another issue for storm spotters in a mobile environment is external vehicle lighting. Of course one thing we must make part of our readiness check list is to make sure all of the vehicle's external lights are in working order; head and tail lights, brake lights, turn signals, and marker lights.

Some storm spotters include on their vehicle an additional flashing light similar to those found on public safety vehicles or public utility vehicles. The reason for this is so that they can be seen at night by other drivers while stationary, perhaps on a road side, to observe weather. There are several problems with this practice. First, use of these type of lights may be prohibited except by emergency vehicles. While storm spotters do provide a valuable service they do not qualify as emergency vehicles. You should first check with local law enforcement or emergency management before using such lights. Second, a flashing light outside the vehicle can have a similar affect as lights inside the vehicle and negatively affect the driver's ability to see. And third, there is no reason a storm spotter should set up stationary in a place where other drivers may have difficulty seeing them. The storm spotter should look for a place that is safe and out of the way of traffic, such as a parking lot. If they must pull off the side of the road they should keep their headlights on and make use of hazard lights.

Safety is our number one priority when storm spotting. If road and weather conditions make it unsafe to spot mobile at night, stay in a safe location. Our ability to see well at night is very limited and under rapidly changing weather conditions it can quickly become too dangerous to be out on the road. If you do storm spot at night don't forget to take a storm spotting partner with you. The driver should focus on driving; let the passenger act as spotter and navigator.

The home-based spotter and net control station also have to be aware of lighting issues. Power outages can quickly put us in the dark. Keeping emergency lighting on hand for such occasions is a must. When we look at the design of our home station or net control station we usually look at providing power back-up for radios and computers. The power back-up may come from a generator, batteries, or a UPS for computers. We should factor emergency lighting into the design. This is usually done for stations that can switch over to generator automatically when the power goes out, such as net control stations operating from an EOC. Not all home-based stations are designed to go to automatic generator back-up when the power goes out. Many home stations may have a generator available or have a battery back-up system. We can save some time fumbling in the dark by keeping a lighting source readily available for our back-up power system.

DIGITAL CAMERAS

In several ways when we make a report of severe weather it is like being the witness to a car accident. The witness is a key part to the understanding of what it happening. He provides the investigator with an account of what happened leading up to an accident and what followed. There are factors that have to be taken into account when interviewing a witness. The mental state of the witness must be considered. He may be under stress or traumatized from the incident. This may influence how he reports on what happened. And frequently witnesses are anxious to talk about what they have seen. This is utilized by police officers as they arrive on the scene. When witnesses are interviewed they are isolated from each other. This allows for an unbiased statement of what was seen and allows the interviewer to compare statements. Often the interviewer can make more progress by communicating to the witness that their assistance may be used to help prevent future accidents. A person is more likely to talk if they feel that their contribution is helpful.

A storm spotter's report is, in many ways, like the witness to the accident. When the storm spotter reports sighting a funnel cloud it begins a series of events that aid meteorologists in putting together a timeline. One spotter reported a rotating wall cloud. Several minutes later another spotter reports a funnel cloud. A few minutes go by and the second spotter reports a tornado on the ground. These reports are coming in with location, time, direction of travel, and being attested to by other reports and radar data. As the information comes into the NWS a timeline of events can be determined. There are

Incoming wall cloud captured on a digital camera. (Larry Brown, K5LMB, photo)

factors though that can influence a spotter's report. Like the accident witness, mental state can be a factor. For example let's say the report of a tornado on the ground came from a spotter that was directly in its path. Of course a spotter should take every precaution to not get in this situation, but it can happen. Being in the path of a tornado will likely affect the mental state of the spotter. And like the witness, spotters are anxious to talk about what they see. That, after all, is why spotters are out there! And again, like the witness, spotters are in essence interviewed in a form of isolation from each other. If I am the only spotter at location X then my report is independent from a report from a spotter at location Y. And in our final comparison with the witness, spotter reports are used to assist efforts to protect life and property. What we report is helpful.

There are other factors, though, that have to be considered when reports are turned in about severe weather. When we are storm spotting how we communicate what we observe is limited by certain factors. The most obvious is experience. We can look at pictures of a wall cloud all day long, but Mother Nature has a tendency to throw curve balls and they don't always look like the pictures. As we gain experience as a spotter and with proper training, another factor, the accuracy of our reports, improves. Stress and anxiety can also play a part in our reporting. When we are under stress our rate of speech naturally increases. This can make communicating effectively more difficult. Our proximity can also influence reporting. Are we looking at a funnel cloud or a wall cloud? If there are trees or structures in the way we may not be able to observe the whole picture. Environmental conditions can also influence reporting. In some parts of the country tornadoes may be rain wrapped, making them difficult to see. But the overall factor in reporting is the individual. We have differing abilities in communicating what we actually see. Some can verbalize what they see with amazing accuracy. 911 operators call this 'painting a mental picture' of an event. Others have some difficulty in verbalizing what they see. And, being human, we are all prone to error.

So how what can we do to help overcome these limiting factors so that our reports are accurate and more valuable? By showing others what we see, along with a report, we can add accuracy and value to our 'witness statement'. Here we are in a different situation from the witness. Rarely does an accident allow enough time for a witness to grab a camera or camcorder. When we go out spotting, though, we have a rough idea of what we're looking for and enough time to bring the tools we need. Let's take a look at how still cameras and video cameras can be useful to the storm spotter and how they can best be used while out spotting. We will also look at how we can share and preserve the images we capture.

In 1975 an engineer at Eastman Kodak built one of the first digital cameras to be used (experimentally). It used a CCD image sensor to capture black and white pictures. The images were stored on a cassette tape and had a resolution of .01 megapixels (MP). One image took 23 seconds to capture. Since then we have come a long way when it comes to capturing digital images. Today most digital cameras in the consumer category (roughly $100-400) range from 5mp to 12mp and at the pro-sumer and professional end they can go up to over 50mp. Even most cell phones are capable of digital images in the 1-3mp range.

Digital photography has revolutionized how we take pictures. It is no longer necessary to wait for film to be developed or wonder if a particular picture came out alright. Today's digital cameras make it possible to get images immediately, check for quality, and even share them with relative ease. For storm spotters a digital camera can be a valuable tool. There are some things about digital cameras that a spotter needs to take into consideration.

First let's look at the zoom feature on a digital camera. The ability to zoom in can help make up for distance from an object or focus in on a particular aspect of an object. Let's say we're looking at what we think is a wall cloud that we are observing from a safe distance. By taking a shot zoomed out we can see the wall cloud in relation to everything else. By zooming in we can get a closer look at the cloud. The quality of the zoomed in picture does depend on a couple factors. First, optical zoom vs. digital zoom and second is stability. Optical zoom relies on the lens of the camera to zoom in on an object. It is just like zoom lenses on film cameras, elements inside the lens move to change the focal length of the lens. The focal length is then measured in millimeters, so a common zoom lens may range from 18-55mm. A digital zoom is not really a true zoom lens. Digital zoom basically crops in on a particular area of the image much like photo editing software does. Once it crops in it enlarges the cropped image to full screen. The trade off is that some image resolution can be lost doing this. When using a digital camera's zoom feature it is better to utilize optical zoom and save the rest for photo editing software. Zoom is given in a magnification measure such as 10x. The camera will have a optical zoom rating and a digital zoom rating. For example a camera may be rated at 12x optical and 4x digital. This would be preferable over a camera that was 4x optical and 4x digital. Stability also is a factor when zooming in on an object. As you zoom in, increasing focal length, any movement in the camera becomes exaggerated. Taking the picture at a focal length of 50mm may not require any stabilization of the camera, but as the focal length increases movement is exaggerated and possibly the amount of light getting into the lens decreases making it necessary to use a tripod or monopod. Some cameras and lenses feature image stabilization that can help keep the image stable.

The next issue we need to look at is memory for the camera. All digital cameras share one feature; they have to have something to store images on. Some have a small amount of built in memory, but generally they rely on some kind of memory card to store images. There are several types of memory cards in use in digital cameras; Compact Flash, SD and Multimedia, XD, Memory Stick, Microdrive, and Smart Media. Compact Flash (CF) cards, which come in two types, are often used in larger digital cameras such as digital SLRs. There are several advantages to CF cards. They are relatively

Device Settings		File Size	Memory card: 2 GB	4 GB	8 GB	16 GB	32 GB
Digital still pictures:	5 MP	1.5 MB	1200	2500	5100	10300	20700
	6 MP	1.7 MB	Approx 1100	2200	4400	8900	17800
	7 MP	2.0 MB	number 900	1800	3600	7300	14600
	8 MP	2.3 MB	of 800	1650	3300	6650	13250
	10 MP	2.9 MB	photos: 650	1300	2600	5200	10350
	12 MP	3.4 MB	550	1100	2200	4400	8850
	~~16 MP~~	~~4.6 MB~~	~~400~~	~~800~~	~~1600~~	~~3250~~	~~6550~~
	21 MP	6.4 MB	300	550	1100	2300	4700

Table 3.2: Capacity of digital cameras' memory cards. (Source: www.lexar.com)

inexpensive and easy to find. And they are designed to have high transfer rates. Currently they are available in capacities ranging from 1 GB to 32 GB. SD cards are also quite common and like CF cards are relatively inexpensive. SD cards have an advantage over CF cards in size. Since they are smaller they can be used in smaller cameras, PDAs, cell phones, and organizers. Like CF cards they are currently available in capacities ranging from 1 GB to 32 GB. XD cards were developed for Fuji and Olympus cameras. They are less widely used than CF and SD and have a smaller storage capacity, currently 1 – 3 GB. Memory Sticks were developed by Sony and are in very limited use. Microdrives, while inexpensive, tend to be unreliable, delicate, and rely on moving parts which require more power from the camera. Smart Media was the forerunner to XD and is being phased out.

Since CF and SD cards are the most widely used and have similar storage capacity the choice between the two really depends more on the camera. Larger digital cameras such as digital SLRs use CF. Smaller digital cameras tend to use SD cards. The thing to keep in mind is image resolution and how much space it takes up on the memory card. As you can see from **Table 3.2**, even at maximum MP and minimum storage capacity it far exceeds the limits of a roll of 35mm film!

The final thing to consider with memory is how it will be transferred from camera to computer or other device. Most cameras come with a cable that allows direct transfer from the camera to the computer without the need of removing the memory card. There are also card readers available that connect to the computer and can read a variety of memory cards. And some computers have this feature built in.

Now let's look at resolution. The best framework to use to understand resolution is by comparing the final digital image with one taken by a 35mm camera. Resolution starts with the camera's rating in megapixels. But since today's cameras are typically rated at 5 MP or higher, resolution is less of a concern unless you are printing poster size images. For example a camera with less than 1 MP capability can capture photo quality images for 4 in by 6 in prints. A 5 MP camera can render photo quality prints up to 8 in by 10 in. And at 10 MP a print 20 in by 30 in can be made at photo quality. Since most of the images a storm spotter is likely to take will be e-mailed or uploaded to share with the NWS and other spotters almost every digital camera on today's market will meet those needs.

Something for the storm spotter to keep in mind is the effect of rain and precipitation on a camera. Most consumer grade cameras do not like being wet. And if water gets on the surface of the lens it can obscure the image. There are commercially available waterproof and water resistant housings available for a wide range of cameras. Also it is possible to use a plastic grocery bag, wrapped around the camera body, to make it semi-water resistant.

A factor that storm spotters may have to take into consideration is low light or night conditions. Trying to take a still image at night can be a serious challenge during good weather, and can be dangerous during severe weather. One way to still get useful images at night or in low light is by using the video capture capability that many digital cameras have. The video feature on a digital camera typically does not offer the same quality as a true video camera, but it can still render useful images. During a tornado outbreak a storm spotter used the video mode on their digital camera to capture about three minutes of storm footage. In most of the video all you see is cars passing on the highway and an occasional flash of lightning. But in one brief flash you could make out

Low-resolution picture of thunderstorm at a distance, taken with camera on a cell phone. (Mike Corey, W5MPC, photo)

View of a tornado at night silhouetted by lightning, captured by the video mode on a digital camera. (Scott Smith photo)

the silhouette of a tornado. By loading the video into video editing software or using basic desktop tools you could take out a still image of the tornado. The image is not going to be high resolution and may be grainy or fuzzy but it is still useful. Besides the drawback of low resolution video mode on most digital cameras is limited by storage space (see **Table 3.2**).

VIDEO CAMERAS

Video cameras are another useful tool for the storm spotter. Since they first came out camera size has reduced dramatically and they can now be carried easily in a pocket or go-kit. There are some things to be taken into consideration when selecting a video camera for storm spotting.

Much like other media, camcorders have several different formats that can be used to store what is captured. Like digital cameras, some camcorders can use the SD and Memory stick formats. There are variations of these formats that are designed to accommodate video. There are also tape formats such as miniDV and Hi8 format. And there are disc formats such as DVD and mini DVD. As with digital cameras, the amount of video you can capture will depend on the storage capacity of the memory, tape or disc.

Camcorder resolution can be a far more complex subject than digital camera resolution. For what a storm spotter needs in video images it comes down to a simple question, "how am I going to present or share this video?" If the video is being e-mailed, uploaded to a Web site, or viewed on a computer or standard television set, most modern camcorders will do. If you are going to watch the video on a high definition television and want the absolute best picture quality, a high definition camcorder would be the way to go. Considering that a storm spotter is simply trying to put an image to what they see an HD camcorder may be more than what is really needed.

The zoom on a video camera is not that different from the zoom on a digital camera, and again there is an optical zoom and a digital zoom. They both work just like they do on a digital camera; optical zoom should be given preference over digital zoom. The difference though is in how much zoom you get on a digital camera. It is not uncommon to find camcorders with optical zoom in excess of 50x and a digital zoom of 2000x. But, *buyer beware*, some manufacturers combine the two zoom measurements to make it look more impressive, ie a 30x optical and 800x digital might be sold as a 24000x.

A big advantage that today's camcorders have is their ability to function well as both a camcorder and a still camera. In many camcorders there is a digital camera or still camera capture mode. Having this integrated into the camcorder allows the spotter to have both digital camera and camcorder in one without sacrificing image quality on either the video or the still image.

Any time we try to capture images in low light or at night it will present problems. For any camera to capture an image it requires light: with less light the ability to capture suffers. For camcorders there are some things to look at when it comes to night and low light situations. First is the *lux* rating of the camera. This indicates how much light the camera needs to capture video. The lower this rating the less light it needs. Second is the night mode feature. Like many digital cameras, some camcorders have a night mode setting that can be used. And finally is the camcorder's format. Some formats are better at recording in low light than others. Of course there are other types of cameras that can be used at night such as night vision cameras, infrared, etc, but these are beyond the scope of storm spotting needs.

Whenever you are active as a storm spotter and decide to take along a camcorder or camera there are some things that must be kept in mind.

First, and always foremost, is safety. Remember you are not out there to be a photographer or videographer, but a storm spotter. If you find yourself more focused on getting 'just the right picture', well pardon the pun, but you've lost focus on what you are there to do. This applies to mobile spotters or fixed spotters. Your first job is to keep yourself safe. If taking a picture distracts you from this it might be better to leave the camera at home. And if you are mobile while spotting do not try to drive and take pictures at the same time. This is dangerous and can get you or someone else injured or killed. If you are mobile you should have someone with you either as driver or spotter, and whoever is not driving should take care of the camera.

Second, sharing the images that are captured. There are many ways that images and video can be shared. Some spotters use Internet techniques such as *Skype* to share images and video with other spotters in a conference call mode. Online sites such as *YouTube* can be used to upload videos for future viewing. Images and videos can also be submitted to the NWS and local media through the Internet. Remember the reason to get a picture or video is so that more details about a weather event can be shared with those who need to know: meteorologists, storm spotters, net control, the media, etc.

And finally, remember that more than storm photos are needed. Getting post storm damage images can also be of great value. A photo of storm damage can help NWS meteorologists determine if the damage came from straight line winds or a tornado. Photos of hurricane damage can give critical information on structure survivability. Don't forget the post-event side of storm spotting.

BINOCULARS

Another handy tool the storm spotter should keep on hand is a pair of binoculars. We may need to get a closer look at some attribute of a storm or other weather event and binoculars can help. So what binoculars should we use?

To answer this question we have to first need to understand some binocular terms. Almost all binoculars are described by a pair of numbers such as 7X50 or 8X30. These two numbers tell us two things about the binoculars, magnification and diameter of the objective lens (the lens furthest from the eyepiece).

The magnification tells us how many times an object will be magnified. A 7X means that an object will appear seven times closer than if viewed with the naked eye. There are also zoom binoculars available that have a range of magnification.

The second number refers to the diameter of the objective lens in millimeters. A 7X50 would have an objective lens that is 50 millimeters in diameter. This factor is also important in choosing a pair of binoculars since it tells you how much light can be gathered into the lens. Since we will most likely be using these binoculars during overcast, cloudy, or possibly night time conditions we will want something that will gather as much light as possible. At the minimum we would likely want to use a pair of binoculars that have an objective lens of 50 millimeters or more.

Some final considerations on choosing binoculars are cost and size. Binoculars can range in price from well under $100 to hundreds or thousands of dollars. And if we are going to use them while in a vehicle (not while driving) we may consider size when choosing a pair.

THE GO-KIT

Most Radio Amateurs that are involved in emergency communications are familiar with the concept of a 'go-kit'. These kits contain essential items used during emergency communications deployments. The kit, which is custom built and stocked by the individual operator, can contain a variety of items for deployments lasting anywhere from a few hours to a few days. The go-kit concept can be valuable to amateurs involved in storm spotting too.

For storm spotters that go mobile when activated a severe weather go-kit can be a valuable and time saving tool. What would go into a severe weather go-kit? There are certain tools a mobile storm spotter is likely to use during most activations, such as field binoculars, small camera, severe weather guide card, flashlight, SKYWARN manual, handheld GPS, manuals for radios and other electronics, spare batteries, and bottled water. What you put in your go-kit is entirely up to you.

Also don't forget to keep your vehicle ready for emergencies. A kit with jumper cables, flashlight, first aid kit, fire extinguisher and other such emergency items should always go with you.

THE SKYWARN WEB SITE

Before we get into general online resources and software for the storm spotter we will first look at setting up a SKYWARN webpage. Almost all of the NWS WFOs have a page on their Web site dedicated to SKYWARN and storm spotting. But at the local level a SKYWARN page can also be set up that addresses local issues, provides information on training and meetings, and provides one location where local spotters can find pertinent weather information or links to information, safety guidelines, the local SKYWARN manual and net control guide, as well as any other information that may be useful. And the SKYWARN page can be linked with the local Amateur Radio club page or ARES / RACES page providing a great recruiting tool for Amateur Radio and serve as a link between Amateur Radio storm spotters and non-Amateur Radio storm spotters.

Brad McConahay, N8QQ, describes the basic steps involved with setting up your local SKYWARN Web page, Web site hosting, domain name, and content in the sidebar on the next pages.

WEB RESOURCES FOR

THE STORM SPOTTER

There is no shortage of weather information available on the Internet. Weather data is available from commercial Web sites, government sources, schools, media, and from individuals that upload data from home weather stations. Data can be numerical (temperature, pressure, wind speed, etc), visual (webcams and still images), radar, satellite, and text.

In this section we will look at some Web resources that are available and useful to the storm spotter. Much of the data available through these sites are free, while some require a paid subscription. We will cover a general overview of what is available and how they can be use full to the storm spotter. We will start with the National Weather Service.

NWS Web sites

The NWS is the federal agency tasked with providing weather forecasts and warnings for the United States and its territories and waters. The data available through the NWS is about as reliable as you can get. There are several NWS Web sites storm spotters should be familiar with, starting with the local Weather Forecast Office (WFO).

The NWS has 122 WFOs throughout the United States and territories.

The local WFO Web sites contain a wealth of information on their particular county warning areas (CWAs). Typically, a

Setting Up the Local SKYWARN Web Site
By Brad McConahay, N8QQ

Web Site Hosting

The traditional method for hosting a Web site is to find a company that offers Web hosting online, and then either install a Content Management System like *Wordpress* or *Drupal*, or learn how to use HTML (Hypertext Markup Language) and build it all from scratch. You can find many companies to host your Web site at a low cost with a simple Web search. It might also be worth checking with your Internet Service Provider as they may include Web hosting as part of your plan. Alternatively, it's becoming easy to sign up for low-cost or free Web-based services that handle almost everything for you. If you have regular updates to share with group members and other site visitors, you might set up a blog site at **wordpress.com.** In addition to blog style updates, it can be used to create a full site with regular Web pages. They include many styles and layouts to choose from that will give you a professional look with very little work.

If you're feeling ambitious and you'd like to try your hand at running a full-fledged social network, you could set up a site at ning.com and treat your members to an experience similar to *Facebook*. The list of these 'Software as a Service' types of hosts is growing every day.

Google deserves a special mention as a company that offers a wide variety of free tools you can take advantage of for hosting parts of your site. You might already be familiar with some of their services, such as *Google* maps, and *gmail* for free e-mail, but they also offer a number of other free tools for the would-be webmaster. *Google Analytics* provides a way to get site statistics regardless of how you host it. There's *Picasa* for storing online photo galleries. With the free version of Google *Apps,* you can provide up to 100 free e-mail addresses for your group's officers or members, shared documents and calendaring, and even facilities to host the entire Web site.

Domain Name

Registering your own domain name is a good idea for several reasons, not the least of which is to provide a short name that's easy for people to remember. You will almost certainly want to switch to a different host for your Web site or e-mail at some point in the future, and by registering your own domain name, you'll always have a permanent address that will be safe to use on your group's printed materials.

You will want to pick a domain name that reflects your group's name or location, and that is as short as possible. The most popular top level domains are .com, .net, and .org, but there's no need to shy away from other top level domains like .info or .us. There aren't too many short domains left available these days, so you might have to be creative. For example, Connecticut SKYWARN uses their state's abbreviation along with the word 'skywarn' to create **ctskywarn.com**. North Alabama SKYWARN uses a combination of abbreviations to keep it really short with **nalsw.net**. National Capital Area SKYWARN uses their repeater's frequency, as in **174300.com**. Oklahoma's **cartercountyskywarn.org** might be a little longer, but is still very descriptive, easy to remember, and a good permanent address.

In addition to providing domain name registration, most domain registrars now offer add-on Web hosting and e-mail solutions at a low cost or free with your domain registration. Most registrars will also provide several methods to point or forward a domain name to wherever you ultimately decide to host it.

What to Put on Your Site

Ideally you will want to break each topic up into its own separate page. If the information is brief, or you don't have the option to create multiple pages, you can also break a single page up into sections. Among the most important information to include is:

● Coverage area. You'll want to be clear about the counties or other areas your particular chapter of SKYWARN covers. It's not uncommon for somebody who lives far away to accidentally request information from you, or submit a membership request, and not realize that they're nowhere near your group's coverage area.

● How to join. If your group is also a club with dues and membership, be sure to include the mailing address for new members to send their payment, the methods of payment you'll accept, and any additional details specific to joining your club. If you have a standard paper form for new members, you can easily scan or convert it to an Acrobat PDF file that people can download and print to mail in. You can also create a Web version of the membership form that people can submit online. An advanced feature you can provide is to accept membership dues online by using *PayPal*. They will take a small percentage of every transaction, but

The home page of W8NWS, the Weather Amateur Radio Network.

it makes the process instant and easy, and it's not too difficult to add to a site.
- Net operations. List the frequency of the repeater used for your on-air nets, including frequencies for any back-up repeaters, and any related procedures that you may have in place. If you have any special access tones for the repeater, you'll want to list that here as well. This is also a good place to mention any protocol information about how your on-air nets operate, such as check-in procedures.
- Reporting criteria. Because different SKYWARN groups have different types and levels of weather conditions useful for their particular area and weather office, you'll want to list the specific criteria your group is seeking during on-air nets. Receiving reports that don't meet the criteria of a SKYWARN net is a common source of frustration, and the Web site is a perfect place to document this for easy reference by all.

Those are some of the most essential pieces of information you'll want to include on your Web site. Here are some extra things you can add to make it even more useful. They may not all apply to your group's situation, but are worth considering:
- E-mail Lists. You may already have an e-mail list, sometimes referred to as a 'reflector', to provide news and information for your members. If you don't, Yahoo Groups is a good way to start one for free, albeit with a bit of advertising. A separate e-mail list can also be handy to use for net activations, with the idea that people will use their cell phone or other mobile device's address to subscribe. This is also an ideal place to add any e-mail list signup procedures, or links related to the lists.
- Social networking. In addition to e-mail lists, social networking sites like *Twitter* and *Facebook* are tools you can use for providing your group's updates or net activation notices. With *Twitter*, you could have one account for the casual dissemination of the group's general news, and then a second account for net activation notices that members can send to their mobiles devices. A *Facebook* fan page can serve as a convenient place for your members to congregate and publicly interact with your group and each other, especially since it provides easy sharing of photos and videos.
- Weather radio information. It can be helpful for members to have a quick reference to NOAA weather radio frequencies and the SAME codes that apply to your coverage area.
- Links. Providing links to related organizations is another nice addition to your site's reference material. You might consider linking to the main NOAA site (**noaa.gov**), your local NWS office's site, other surrounding SKYWARN chapters, local amateur radio clubs, and the ARRL or other organizations your group might be affiliated with.
- Event calendar. If you have a lot of planned events within your organization or other dates that are important for your members, you could add an event calendar to your site. One of the easiest ways to do this is to use a *Gmail* or *Google Apps* account to set up and embed a calendar on your site. This method also allows you to set permissions for other authorized people to make updates.
- A Photo Album. If you have pictures from trainings or other events, you can start a photo album and either link it from the site, or embed it. Using **Flickr.com** or *Google's Picasa* are both popular ways to do this.
- How to become an Amateur Radio operator. People will run across the site who are interested in becoming a SKYWARN spotter but who don't know much, if anything, about Amateur Radio. For this reason, it's a good idea to link to resources for question pools, tutorials, and testing dates and locations. Many of these things can be found on the ARRL Web site.
- Last but not least, don't forget to give people and organizations credit where credit is due. The trustee or sponsor of the net repeater, club officers, and net control operators are all people who probably deserve a mention. And while you're at it, don't forget to give yourself credit for all the hard work you've put into your Web site!

Storm Spotting & Amateur Radio

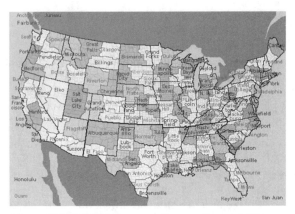

Map of NWS WFOs across the continental US. (National Weather Service)

NWS Western Region home page. (National Weather Service)

WFO's Web site will feature meteorological information on current hazards, current conditions, radar imagery, forecasts, and climate. Information on weather safety and education and outreach are also available. Most WFOs also have a page dedicated to SKYWARN, Storm Ready Communities, and general preparedness guides.

Usually a map appears when you log on to the local WFO home page showing the CWA and any watches, warnings, advisories, short term forecasts, or hazardous weather outlooks. By clicking on your county you can bring up a page showing current conditions, the short-term forecast, links to text of short-term forecasts or hazardous weather outlooks, and radar and satellite data links.

Another important feature available on most WFO Web sites is a link to submit storm reports. This page usually has drop down menus with set terms to describe weather events. This helps prevent any confusion when describing what is going on, for example "marble size hail" can range in size while "nickel size hail" has an exact size. Along with ability to submit storm reports many WFO Web sites include information on submitting storm video or photos and provide access to submitted storm reports.

One of the best things a storm spotter can do is to get familiar with the Web site of their local WFO. It should be a daily practice to log in to it and check the current forecast and any weather hazards that might be expected.

The next step above the local WFO is the regional office. There are six NWS regional offices; Central, Eastern, Southern, Western, Alaskan, and Pacific. The regional office provides weather information on their particular region of the country. The regional Web site also serves as portal to national weather information. Regional offices also have available regional publications on weather data and regional weathers programs.

And at the national level is the main NWS Web site.[11] The national Web site provides information on weather forecasts, hazards, warnings, and observations across the country. There is also information available on NWS sponsored programs such as SKYWARN. If you cannot find what you are looking for at your local WFO Web site then check in to the national NWS site.

Specialized NWS Offices

The Storm Prediction Center (SPC) located in Norman, Oklahoma provides short-term weather forecasts to the public and the NWS WFOs. The SPC is responsible for issuing tornado and severe thunderstorm watches for the lower 48 states and severe weather outlooks. The SPC Web site[12] provides access to several different products; overview of severe weather across the US, convective outlooks, watches, mesoscale discussions, warnings and advisories, storm reports, mesoscale analysis, and fire weather information. The SPC Web site also provides access to research and forecast tools.

For the storm spotter the SPC is a valuable source for seeing what the weather picture is like beyond the local level and what weather hazards may be developing. The storm reports the SPC collects, which come from local WFOs, also provide a great recap of severe weather events. The SPC homepage also provides weather related news articles.

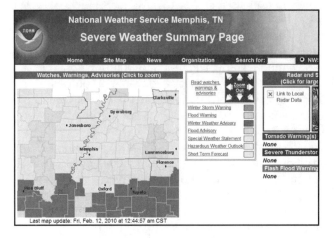

Home page of NWS Memphis WFO. (National Weather Service)

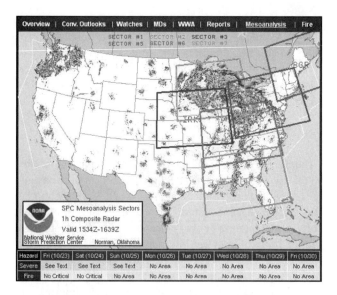

Storm Prediction Center's mesoanalysis. (National Weather Service)

For hurricane and tropical weather there are two NWS Internet resources; the National Hurricane Center (NHC) located in Miami, Florida and the Central Pacific Hurricane Center (CPHC) located in Honolulu, Hawaii. While both these centers have similar tasks there are some subtle differences. Let's look at each one's mission, available products, and coverage area.

The mission of the NHC is to issue watches, warnings, forecasts, and analysis of hazardous tropical weather and to conduct research. There are several different watches, warnings, and advisories that the NHC may issue:

- *Gale Warning* - A warning of one-minute sustained surface winds in the range 34 knots (39 MPH or 63 km/hr) to 47 knots (54 MPH or 87 km/hr) inclusive, either predicted or occurring and not directly associated with tropical cyclones.
- *High Wind Warning* - A high wind warning is defined as one-minute average surface winds of 35 knots (40 MPH or 64 km/hr) or greater lasting for 1 hour or longer, or winds gusting to 50 knots (58 MPH or 93 km/hr) or greater regardless of duration that are either expected or observed over land.
- *Hurricane Warning* - A warning that sustained winds 64 knots (74 MPH or 119 km/hr) or higher associated with a hurricane are expected in a specified coastal area in 24 hours or less. A hurricane warning can remain in effect when dangerously high water or a combination of dangerously high water and exceptionally high waves continue, even though winds may be less than hurricane force.
- *Hurricane Watch* - An announcement for specific coastal areas that hurricane conditions are possible within 36 hours.
- *Storm Warning* - A warning of one-minute sustained surface winds of 48 knots (55 MPH or 88 km/hr) or greater, either predicted or occurring, not directly associated with tropical cyclones.
- *Tropical Storm Warning* - A warning that sustained winds within the range of 34 to 63 knots (39 to 73 MPH or 63 to 118 km/hr) associated with a tropical cyclone are expected in a specified coastal area within 36 hours or less.
- *Tropical Storm Watch* - An announcement for specific coastal areas that tropical storm conditions are possible within 48 hours.

Beyond issuing watches, warning, and advisories, the NHC offers several products on their Web site valuable to storm spotters in areas affected by hurricanes. From the main page information can be accessed on current areas of interest for development of severe tropical weather. Satellite and radar imagery is accessible as well as text data. Also accessible are marine forecasts. Information can also be found on the Web site on hurricane preparedness and historical data on storms.

The NHC's coverage area is the North Atlantic and Eastern Pacific areas. This covers most of the continental United States, Mexico, Central American, the Caribbean, and the North Atlantic Ocean, including the area where big hurricanes develop, the west coast of Africa near the Canary Islands.

The CPHC, like the NHC, is tasked with issuing watches, warnings, advisories, forecasts, discussions, and statements on tropical cyclones in their coverage area. The CPHC is activated by the WFO Honolulu when certain criteria are met; 1) a tropical cyclone moves into the Central Pacific from the Eastern Pacific, 2) a tropical cyclone forms in the Central Pacific, or 3) a tropical cyclone moves into the Central Pacific from the west.

The watches and warning terms are similar to those issued by the NHC, but there are some differences in the definitions for some terms:

- *Hurricane Warning* - A warning that sustained winds 64 knots (74 MPH or 119 km/hr) or higher associated with a hurricane are expected in a specified coastal area in 36 hours or less. A hurricane warning can remain in effect when dangerously high water or a combination of dangerously high water and exceptionally high waves continue, even though winds may be less than hurricane force.
- *Hurricane Watch* - An announcement for specific coastal areas that hurricane conditions are possible within 48 hours.
- *Tropical Storm Warning* - A warning that sustained winds within the range of 34 to 63 knots (39 to 73 MPH or 63 to 118 km/hr) associated with a tropical cyclone are expected in a specified coastal area within 36 hours or less.
- *Tropical Storm Watch* - An announcement for specific coastal areas that tropical storm conditions are possible within 48 hours.

Like the NHC, the CPHC offers several products that are valuable when tropical cyclones threaten. From the main page we can access information about currently active storms. Also available are radar, satellite, and text data on tropical weather. Sea surface temperatures and forecast information is also available. Tropical cyclone history and preparedness information is also available.

The coverage area for the CPHC is the central Pacific Ocean area from 140° West longitude to the International Date Line and from the Equator to 30° North Latitude. This covers nearly five million square miles.

NWS *ESpotter*

To help make communication between local storm spotters and local NWS offices more efficient the NWS has developed the *ESpotter* program. The program is a Web site that local emergency managers and storm spotters can access to submit local storm reports. Users must sign up for an account to access the system and can do so at no charge.

Once logged on users are provided with a national map showing watch / warning / advisory areas. By clicking on an area the user is taken to that locations local NWS office. The Web site also allows messages to be sent between the NWS office and the user.

The main function of *Espotter*, though, is to provide an efficient way for storm spotters to submit reports directly to the local NWS office. From the main page the user can click on 'create report' and be taken to the report form. The form covers the basics; tornado, funnel cloud, wall cloud, hail (including size), high wind (estimated speeds), flood, flash flood, and other. It also allows the user to indicate if any damage or injuries relating to the event have occurred. Finally it allows the user to provide additional details relating to the event.

Interactive NWS

Another valuable program available from the NWS is *interactive NWS* (*iNWS*). This Web site allows registered users to have mobile access to several NWS products.

iNWS Alert sends text message and e-mail severe weather alerts to users. A user may configure it to only send certain alerts, ie tornado warnings and thunderstorm warnings to specific locations.

AHPS Mobile gives users mobile access to hydrographs, current and forecasted stages, and river impacts. 'River Watch Points' can also be set up to monitor local points of interest. This feature is handy for areas prone to river flooding.

iNWS Mobile is a Java-based interface that runs on mobile devices and provides access to weather service watches, warning and advisories, radar and satellite imagery, observation, and point forecasts. This feature may not work on all mobile phones. *iCWSU* is similar to *iNWS Mobile* but is geared for the aviation community. It provides information on hub forecasts and discussions, TAFs and METARs, storm summaries, convective and tropical outlooks, aviation hazard graphics and imagery.

iNWS Mobile Web is a mobile Web application that allows users to access commonly needed NWS products such as weather information by zip code or city, current watches, warning and advisories, current conditions, radar and satellite feeds, and National Weather Service forecasts.

WEATHER WEB SITES

There are several commercial Web sites that provide reliable weather information. They do not require the user to pay any fees, download software, or subscribe to a service, although they may have some features that require a subscription fee, software, or signing up for an account. We will consider the pros and cons of both free information and subscription services. In this section we will look at a few of these and some of the features that may be useful to a storm spotter. Almost all of these sites provide some of the same basic information; radar and satellite data, watches and warnings issued by the NWS, local conditions, and local forecast information. And most also offer a mobile service as well. Instead of looking at basic common features we will look at some of the unique features these Web sites offer. Of course over time features come and go, so the only true way of seeing what each site offers that is unique is to log in and explore the site. Which site or sites you select to go to for weather information is up to you.

AccuWeather

AccuWeather[13] was founded in 1962 by Dr Joel Myers and is headquartered in State College, Pennsylvania. Their online service AccuWeather.com provides weather forecasts for the entire United States and over two million worldwide locations.[14] Originally the service provided weather forecasts to utility companies, but in 1971 was expanded to provide weather forecast information to media. Today they provide weather forecast information to media, Internet users, news-

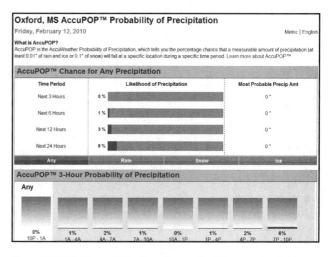

AccuPOP™. (AccuWeather ® used with permission)

papers, wireless customers, and government clients.

Through the AccuWeather.com online site, users can access several different weather products including forecasts, radar, maps, video, news, and extreme weather. Access to these items are common for most weather Web sites, however, there are features to each that are unique to AccuWeather. Let's look at some of the features unique to AccuWeather.

If we look under the forecast features we will find Accu-POP™. AccuPOP is a feature that provides the probability of measurable precipitation over a given period of time. This feature can provide probability of precipitation of rain, snow, or ice over a 96 hour forecast period for any location in the United States. This can be a valuable tool for storm spotters. On a day to day basis it can be used for determining the likelihood of precipitation. And during periods of heavy rain, ice, or snow it can give us an idea on what to expect over the following hours or days.

Under the radar category there is another useful feature called MapSpace. This combines radar data with a dynamic map similar to maps found on *Google Earth*. The user can manipulate the maps in several ways; drag to a new location, enter specific location information, change weather data displayed on map, change transparency levels, zoom in and out, and animate the map.

Many key features including AccuPOP and MapSource are available for no charge. AccuWeather.com does also have a paid subscription service, AccuWeather.com Premium, that offers users more information. There is also a professional level paid service, but for our purposes we will only look at the Premium service.

The first noticeable difference between the free service and the paid service is the lack of advertising in the paid service. Many sites, both weather and non-weather, can provide data for free because of the advertising space they sell. By paying a subscription fee users can see only the pertinent data without advertisements. There are features beyond ad-free data for paid users. Paid users can also use Storm Timer and Table, which provides timeline information on severe weather. This can be useful for tracking incoming storms and planning storm spotter activation. Additional features of the paid service include; historical data, forecast models, customizable features including a homepage with local weather data, and a planner feature. As of this writing the monthly subscription fee is $7.95.

WeatherBug

WeatherBug[15] is a little different than most online weather services. While features such as local forecast and severe weather information are available on the WeatherBug Web site, the strength of WeatherBug is local weather observation. WeatherBug uses data from over 8000 tracking stations and 1000 cameras to provide real-time weather observation data. This data is delivered to users through a downloadable desktop application.

WeatherBug started in 1992 as a way of bringing meteorology into schools. Schools could be networked into a

WeatherBug desktop application. (WeatherBug image)

system of weather observation data and were also provided in class curriculum to use. The schools were later linked to the broadcast meteorology community through partnerships with local network affiliates. In 2000 the desktop application was launched. This made it possible for a broad range of users to access weather observation data as well as forecasts and weather alerts.

The WeatherBug desktop application provides basic weather data; local forecast, severe weather alerts, radar, etc. Unlike other services this is done in a separate application that does not require a Web browser. It also allows the user to access local cameras which can provide a visual cue on weather conditions.

Another handy feature that WeatherBug offers is a mobile application for cell phones. Only certain cell phones are supported. The service provides WeatherBug data to mobile users. There are two types of service available; WeatherBug (free) and WeatherBug Elite (paid service). The paid service, like others, offers users an ad-free option.

Stormpulse

Stormpulse[16] is a fairly new weather Web site. It was started in 2004 and is a great source for tropical weather information.

Stormpulse (used with permission).

The main page provides current weather data on the Atlantic and Pacific basins, satellite imagery, and a storm archive. Recently a feature was added that provided severe weather information for the United States.

The data for the site comes from a wide range of reputable sources. Storm track and forecasted path data come from the National Hurricane Center. Cloud cover imagery comes from the NERC Satellite Station at the University of Dundee. Base imagery is provided by NASA. And forecast models data are compiled and retrieved from the South Florida Water Management District.

Storm spotters, particularly those in hurricane prone areas, can benefit from the data provided on Stormpulse in several ways. In real time we can immediately see where the storm is located and a time line on where the storm is likely to track. Data such as current wind speed, pressure, movement, and category are also provided. On the map we can overlay a variety of data to give us a more robust picture on what the storm is doing and what it might do; forecast models, radar data, and cloud cover can be applied. We can also apply historical tracks, map grid, ocean buoys, wind fields, and map labels. These features allow the storm spotter to follow the storm and make necessary preparations.

The archive data can also be a useful tool for storm spotters. After a hurricane our memory of what actually occurred often becomes a little less than '20/20'. And as time moves on exact details may be forgotten. By accessing historical data we can get snapshots of storm data along its path. If we combine this with our logs, records, and after action reports we can easily build a solid local archive of the storm and how we responded to it. Additionally this gives us the tools to compare one storm to another. We can compare the storm's impact and how we responded to each.

Another useful feature is Stormpulse API. This feature allows hurricane tracking maps to be embedded in a Web page. This can be handy for SKYWARN groups in coastal areas prone to hurricanes and tropical storms.

Stormpulse is freely available and offers users an amazing array of features. Like other sites they also offer paid, ad-free options too. At the Pro level ($3.95 per month) users gain several additional features; 12-hour satellite loops, auto refresh, tropical watches and warnings, and the ability to save your view. There is also a Group level (starting at $24 per month) which adds the abilities of multiple users and secure data sharing.

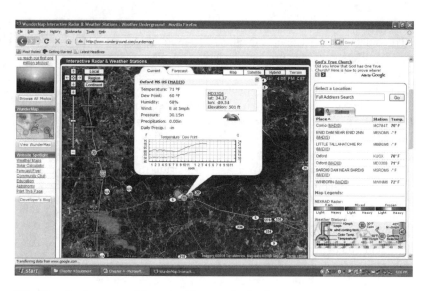

Weather Underground (used with permission).

Weather Underground

Like other free Internet weather sites Weather Underground[17] traces its roots to an effort to help bring weather into the classroom. Weather Underground started at the University of Michigan in 1991 as a telnet-based system for real-time weather information. Over time the number of users and information available grew substantially. By 1995 Weather Underground became a separate entity from the University and was providing real-time weather data for 550 US cities. Today Weather Underground provides a wide range of weather information to a wide range of users. The Web site features access to several standard sources of weather data; local forecasts, radar and satellite data, tropical storm information, information about specific weather events such as hurricanes and tornadoes, and historical information. Weather Underground also features a page where users can access NOAA Weather Radio and listen online.

There are a couple features unique to Weather Underground that are worth noting; local weather stations and the WunderMap.

Weather Underground is not alone in providing access to local weather stations. But one handy feature they do provide is a guide on setting up a weather station. The Web site guides you through choosing a weather station, placing it for accurate measurements, installing and configuring software, and uploading data to the Internet. Registered users of Weather Underground can upload data from their weather station and have it accessible online on the Weather Underground Web site. Weather Underground does not only show local weather stations that are registered with them. You can still access data from other local weather stations such as MADIS and CWOP (Citizen Weather Observation Program) stations.

WunderMap is a feature that gives the user control over a dynamic map that can access a wide range of weather information. The user can set the map, which is *Google Earth* based, to view NEXRAD radar, local weather stations, severe weather, Webcams, satellite data, United States Geological Survey river data, model data, hurricane, fire, and tornado information, National Digital Forecast Database information, and even photos of the local area. The map also allows the user to select from a standard map, satellite image, hybrid map, or terrain map and has zoom controls.

The Weather Channel

Most people are familiar with The Weather Channel (TWC)[18] and the services that they provide on TV and online. TWC's Web site is another great source for weather information for the storm spotter. From their Web site we can get local forecasts, severe weather information, current weather conditions, etc. information widely available from several reputable sources. But also available online and through television programming is a wide range of educational programs about weather. In the training section we will look at this in more detail.

COMPARING SOURCES OF WEATHER INFORMATION

There is no shortage of sources freely available online as well as through TV and commercial radio that can provide the storm spotter with weather information. Is one any better than the other? No one can say with absolute certainty and that is definitely not the goal here. Our goal is to get the storm spotter to look at a range of sources. They may find one that they prefer or they may like several. If you're looking for an online source for weather information there are plenty of places to turn to. For day-to-day stuff almost any of these source, and many others (we didn't even cover local TV station Web sites), can provide you with the local forecast and current conditions. As storm spotters we have different needs when severe weather threatens. We need up to the minute radar information, timely delivery of watches, warnings, and advisories, and accurate short term forecasts.

Radar imagery from most sources is going to be current. As long as your Web browser is refreshed regularly you should have a current image of what is going on. All watches, warnings, and advisories come from the NWS, so no matter where you hear it from it originally came from the NWS. There is almost always a time delay from the issuance of watches, warnings, and advisories and when they appear on the Internet. To compensate for this delay it is a better to use NOAA Weather Radio to receive these. There is less lag time when you get it direct from the source. And something to keep in mind with short term forecasts: many online sources deal with weather information at national and international levels. There will be less coverage about what is happening now and in the next few hours through these sites than through local sources such as the local NWS office and local broadcast meteorologists.

SOFTWARE

Besides weather Web sites there are also several computer programs that are of interest to storm spotters. In this section we will look at a few of these programs. We will consider several factors; functions and features, benefits for the storm spotter, and cost. All of these programs rely on an Internet connection. This is important because we need to have a connection to servers and other sources that provide the data we're looking for. And some programs also have a communications function to them that allows us to connect with other spotters and the NWS. All the programs basically are stand alone programs that operate separate from your Web browser, although some have both a stand-alone component and an online, Web site component.

The Spotter Network

The Spotter Network (SN)[19] is a product of Allison House LLC and has been around since early 2006. The idea behind the SN was to bring together storm spotters, storm chasers, coordinators, and public servants to provide highly accurate severe weather reports. It also helps to improve the communications flow of information to NWS offices and severe weather researchers.

Like technologies such as *EchoLink* and *Skype* the SN also provides a tool storm spotters can use to help fill communication gaps. All that is required to use the SN is an Internet connection. Registered users (registration is free) can submit reports online at the SN Web site or via a downloadable graphical user interface (GUI). Reports submitted through the SN are then relayed to the local NWS office via ESpotter. Typically a storm report submitted on the SN arrives at a NWS office via ESpotter in less than 45 seconds.

Two factors that make the SN valuable to NWS offices and the users of the network are accuracy of reporting and location of spotters. Accuracy is ensured through training required for all registered users. The training, which is not the same as SKYWARN training, is offered online and covers weather, severe weather, and reporting criteria. Location is provided by the user by manually entering their location and coordinates or by an interface to a GPS receiver. Beyond accuracy of reports and location the SN also utilizes a thorough quality review process. An advisory committee conducts after the fact reviews to ensure quality of reports and false reports are never tolerated.

The SN also provides data to users other than the NWS. Data from the SN is shared with weather software packages

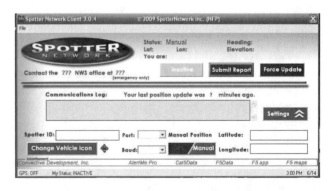

Spotter Network user interface. (Spotter Network, used with permission)

such as GRLevelX and StormLab, *Google Earth*, and for RSS feeds. Information is also shared with severe weather researchers.

The SN is not a product of the NWS or is endorsed by the NWS. It is the product of a private, non-profit organization. Since its beginning though storm spotters, researchers, the NWS, and others have acknowledges the valuable service that the SN provides.

WeatherTAP

WeatherTAP is a commercial weather company that provides, to subscribers, access to a stand-alone radar tool and a Web site[20] where additional weather information may be obtained. Subscription to their services does have a month-ly cost of $7.95 and users have the option of annual sub-scriptions.

Like most sites there are some unique features to WeatherTAP, such as RadarLab. Radar is one of the most valuable tools storm spotters can utilize. It gives us the ability to see what is coming, what is developing, and where weather threats may be present. Many weather sites and software programs give us base reflectivity and composite reflectivity radar views. These views can be put in motion to give a more complete picture. Additionally we can usually access storm velocity and Vertically Integrated Liquid (VIL).

RadarLab is a stand-alone Java-based program that gives users access to radar data as well as additional weather information. Users may customize the program for their location and to display certain information on start up.

The radar data available includes base reflective view at several different radar tilts and composite reflective view. Radial velocity and storm velocity data are also available at different tilts. Information on echo tops, precipitation totals, and VIL is also available. The program also allows the user to overlay data on the radar map such as GPS location, place names, roads and highways, and borders.

A handy feature to this program is the amount of custom data that can be applied to the radar map. Users have the option of including on the map current temperatures, lightning information, watches / warnings / advisories, storm tracks, local storm reports, data from Spotter Network, and surface observations. Beyond the amount of data that can be accessed or applied the program also allows users to animate, zoom in, change map views and radar sites, and capture still images from the radar screen.

RadarLab HD is also relatively easy and straightforward to use. It does not take long to download, set up, and customize to meet your needs. There are also other versions of RadarLab, Classic and HD with GPS. The Web site also keeps users updated with news on new features and other information.

Let's take a look at some of the other features that may be useful to storm spotters on WeatherTAP's Web site.

Users can access forecasts, severe weather information, information for mobile users, and radar and satellite data.

Another unique feature is aviation weather information, with access to Terminal Doppler Weather Radar (TDWR). This radar system is used at airports across the country and adds to the number of radar sites users can access. TDWR is a higher resolution radar image (at ranges close in to the radar). Since these sites are at airports near major cities it gives users a higher-resolution image at a city level. There are some advantages and potential disadvantages to using TDWR. A primary function of TDWR is to alert airports of the presence of wind shear, an item of interest to storm spotters. There are also times where NEXRAD radar may be down, TDWR can be used as a back-up so you don't lose local radar imagery. TDWR also is designed to greatly reduce ground clutter. This can be an advantage but it can also reduce nearby precipitation that may be of interest.

WeatherTAP also allows users to access aviation forecasts and current conditions. We must remember that the aviation community's interest in weather extends beyond what is happening at the ground level. They also have a keen interest in what is happening in the atmosphere. By looking at upper level conditions we can get a more complete picture of what is going on with the weather.

Another unique feature that is becoming more common amongst weather Web sites is KML file access.

Users of WeatherTAP can download KML files for *Google Earth*. These files provide WeatherTAP features in *Google Earth*.

And, finally, WeatherTAP also provides online guides to the range of products. So if you find yourself stumped between base reflectivity and composite reflectivity you can access an online tutorial to help.

WXSpots

WXSpots[21] is a weather program offered by Scott Davis, N3FJP. It is a stand-alone program that is available free to anyone interested. Once downloaded you create an account and log in to a server. Once logged in you may submit weather reports or you may connect your home weather station to the program to submit weather reports automatically. The software features a chat feature too so all connected users can stay in touch.

WXSpots offers some unique features that storm spotters will find quite handy. First is that it is available to anyone, whether or not they are an Amateur Radio operator. This can help bridge the gap between Amateur Radio storm spotters and non-Amateur Radio storm spotters and provide a forum to communicate in real time, share weather information, and coordinate severe weather response. The ability for users to chat works in several ways. A user may connect with an individual user, local group, or all connected users. WXSpots also can pro-vide some archival data to users. You can log in and see recent weather reports submitted by users. The interface also allows users to monitor radar and severe weather data.

There are some critical points that must be made about this software, though. First is that a user cannot assume that just because a storm report was submitted through WXSpots that it was received by the local NWS office. This software is for sharing weather observations, whether or not those observations meet the criteria of severe weather. Users may submit routine weather conditions or severe weather conditions. Local storm spotters should still continue to submit reports via their local NWS office, ESpotter, or through Spotter Network.

Weather Defender

Weather Defender is a product from Swift Weather and emerged from another weather product that they offered, Swift WX. Weather Defender is a stand alone, Internet-based, weather data interface. It is a subscription service that starts, currently, at $29.95 per month. Subscribers may pay monthly or one year at a time. There is also a commercial option subscription available. The commercial plan expands some features and is more suited for businesses and public safety agencies.

Certain features are common to most weather software, including Weather Defender; users have access to radar data, watches, warnings, and advisories, current conditions, and forecasts. However, Weather Defender also offers several unique features that are valuable to storm spotters. The software allows the user to establish an alert perimeter (up to three in the residential version) that will sound an alarm when severe weather breaches the perimeter. Many programs can do this based on county areas, but Weather Defender allows the user to draw a circle around the area they want to monitor, regardless of county boundaries. When a alarm is generated the software can notify the user by an alert screen or alarm sound or be set up to send them via e-mail or text message.

The software also provides the user with several seasonal, industry or activity specific weather maps. Users can access custom summer and winter weather maps. There are also maps for marine, aviation, and ranger use (wild fires and earthquakes). And for Amateur Radio there is a custom map that shows APRS data, repeater towers, and local storm reports.

Another unique feature is that the user may layer data on to the map. The layers are highly customizable and can be placed in any order that the user wishes to place them in.

Other features of Weather Defender include lightning data, GPS and GIS capabilities, national radar and satellite imagery, and watch, warning, advisory data.

GRLevelX

GRLevelX[22] from Gibson Ridge offers three programs that are of interest to the storm spotter or anyone interested in weather who wants a high configurable, feature rich stand-alone program to monitor weather data: GRLevel2, GRLevel3, and GR2Analyst. Unlike most programs we've discussed these software packages are purchased by the user, there are no subscription fees. Let's look at the each program and what it provides. GRLevel2 is a *Windows* based program that allows the user to view NEXRAD II data. The user can monitor base reflective and velocity, storm relative velocity, and spectrum width sweeps. The software also allows the user to pan, zoom, put radar in motion, and capture screen images. Watch / warning / advisory information is also accessible. The program can also be integrated with a GPS receiver. You can download a fully functional trial of the program from the Web site. The current purchase price is $79.95.

GRLevel3 is similar to GRLevel2 but offers users access to NEXRAD level III radar imagery. It also adds a few extra features in the radar display; composite reflectivity, echo tops, vertically integrated liquid, storm attributes, and rainfall amounts. Like GRLevel2 a fully functional trial can be downloaded from the Web site. The current price is $79.95.

The next step up is GR2Analyst. At first glance GR2Analyst looks a lot like the GUI for Level 2 and 3, but it offers significantly more features. GR2Analyst offers not only standard two dimensional radar imagery, but three dimensional radar modeling of storms. This allows the user to see not only a top down view of the storm, but a complete picture of the storm's structure. This is called the volumetric display. The volumetric display can give two different views of a storm. The user can take a thunderstorm cell, look at it from just about any angle above ground level, and see the structure of the storm. The user may also, by adjusting settings, peal away parts of the storm so they can virtually see inside. The user may also pull out cross sections of the storm for analysis.

Like the previous two GRLevelX products the user also has access to a wide range of radar and weather data; base reflectivity and velocity, storm relative velocity, echo tops, vertically integrated liquid, and VIL density. It also gives the user access to maximum expected hail size, probability of severe hail, and normalized rotation. Watch / warning / advisory polygons can also be displayed. The radar data displayed on GR2Analyst comes from NEXRAD level II sites.

Screenshot from Weather Defender showing radar data plus repeaters and APRS. (Weather Defender, used with permission)

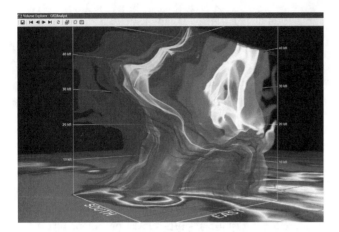

Volumetric display of May 1999 Moore, Oklahoma tornado from GRLevel 2. (Image provided by Michael Laca)

Like the other two programs you can download a trial version of the software. The current price of the GR2Analyst package is $250. For all of the GRLevelX products the user may add additional customizations by adding place file into the program. These place files are available from several sources and can add a variety of new data to the display such as METAR observations, APRS data, SPC watches and outlooks, and local storm reports. Also all of these programs offer GIS support.

Allison House

A useful add-on for any of the GRLevelX programs or StormLab is a data feed from Allison House. There are several options, at different prices, available: Storm Watcher, Storm Chaser, Storm Analyst, and Storm Hunter. Generally each product offers the same data feeds; surface METARs, offshore surface observations, CWOP (Citizen Weather Observation Program – see below) data, APRS data, GOES satellite images, etc. The difference is in the types of NEXRAD radar site that they support. Storm Watcher offers no data feed from NEXRAD sites, Storm Chaser offers data from NEXRAD level 3 only, Storm Analyst offers data from NEXRAD level 2 and NEXRAD level 2 super-res, and Storm Hunter offers data from all three NEXRAD sites.

To use one of these data feeds simply select which data you would like to add. You will be given a URL address to copy and paste. Copy the address and paste it into the place file folder in your weather software (GRLevelX or StormLab). This will add that data to the display.

It should be noted that while data is available through subscriptions services such as Allison House, there are also data available for free from a variety of user support groups online.

StormLab and Interwarn

StormLab and Interwarn are both products of StormAlert Inc. These two programs serve different functions and can be purchased separately or as a package. StormLab is a stand-alone program for viewing radar and other weather data. Interwarn is a program that is designed to monitor weather watches, warning, and advisories. Each program can be customized by the user.

StormLab provides the user with access to over 150 radar sites across the United States and about two dozen different radar products including; reflectivity, velocity, storm relative velocity, wind profiles, storm cell data, and precipitation

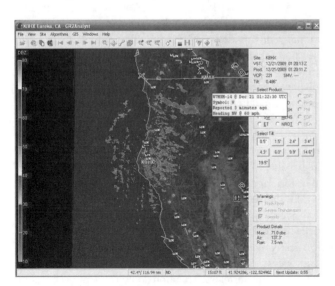

CWOP and APRS data displayed on GRLevel 2. (Used with permission)

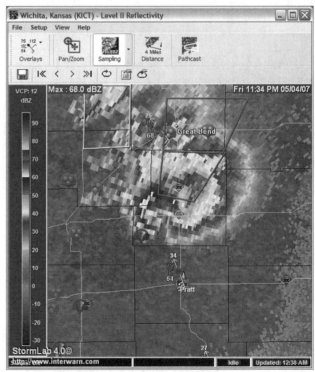

Radar image of Greensburg, Kansas tornado overlaid with warnings, storm reports, live spotters, and more. (Source StormAlert Inc, used with permission)

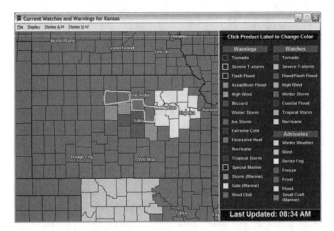

Current watches and warnings view in Interwarn. (StormAlert Inc, used with permission)

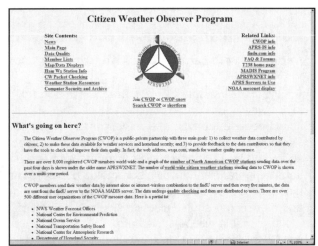

CWOP Web site.

estimates. It also allows the user to integrate GIS data and GPS location data. The user also has control over several viewing options; pan, zoom, animate, and viewing past data. StormLab comes in two forms; regular and 'supercharged'. The supercharged version gives the user access to advanced features such as GPS, customization with data from Allison House, METAR data, and distance tools.

Interwarn is a stand-alone program that provides access to watch, warning, and advisory data as well as text data from the NWS. The user configures the software for their location and for the information they wish to receive. The data can be displayed on the screen, sent as an e-mail or text message, or printed. It is like having a personal weather wire service.

The user interface, once customized, offers the user a variety of instant information. The initial screen provides the user with any bulletins for their specified area. And by clicking on the Real Time Watch / Warning Map display the user can see any watches, warning, or advisories for their state, a very handy feature for keeping up with what is going on around you. The next tab provides local and national storm reports. The third tab provides current conditions. This data is organized by state and city. Your exact location may not be included, but you should be able to find a nearby city. The next tab is for retrieving text data from the NWS. This can include forecasts, observations, current conditions, etc. The user can select the exact text products they want to retrieve. Next is the tab that accesses graphical data. Here the user can access forecast information, satellite observations, and Doppler radar data. Finally is a tab for quick links. The software comes with some pre-loaded links to weather and news data and also has spots for user defined links.

CWOP

CWOP stands for the Citizen Weather Observer Program[23]. It is a private-public partnership between individual citizens who collect weather data on the one hand, and the weather services and homeland security on the other. The program has three objectives: 1) to collect weather data contributed by citizens; 2) to make these data available for weather services and homeland security; and 3) to provide feedback to the data contributors so that they have the tools to check and improve their data quality. The users of this data range from other citizens to universities to government agencies to commercial weather companies. There are over 8000 registered CWOP members around the world. The data is sent via the Internet to FindU.com and from there to the NOAA MADIS (Meteorological Assimilation Data Ingest System). To access data from any CWOP station all that is required is an Internet connection to a site that displays the data such as Weather Underground or weather software such as GRLevelX.

The data from CWOP stations can be a valuable tool for local storm spotters. By checking the data coming from your local CWOP station you can see temperature, pressure, and precipitation. Graphs allow us to also see trends over the last several hours or days. All of this data is stored online so we can also go back and retrieve past data.

The CWOP station consists of four components. First is a weather station that is properly set up and calibrated. Second is a computer that can connect to the weather station. Most weather station manufacturers have cables available for this. Third is software that can gather the data from the station and configure it into a data packet to send to the server. Software is available from weather station manufacturers as well as independent sources. And finally is an Internet connection so the data can be uploaded.

As data is uploaded to the Internet it is checked for accuracy. You can sign up to receive daily reports on any errors coming from your CWOP station.

HOME WEATHER STATIONS

For most Radio Amateurs involved in SKYWARN and storm spotting our interest in weather goes beyond just severe weather and SKYWARN activations. We generally check the forecast each day. We like to know what is happening now, even when it is clear, warm, and sunny. Our family, friends, and colleagues tend to turn to us to find out what

Best Weather Stations - Factors to Consider When Making Your Purchase

(reprinted with permission of WeatherShack.com)

- *Weather Variables.* This is the most important factor when purchasing commercial weather stations because the type of weather measurements (variables) you wish to monitor will determine if you should be looking for a basic personal weather station (temperature, humidity, barometric pressure) or a complete professional weather station (wind, rain, and more). You should also consider how a unit's indoor console displays this data.
- *Cost.* How much you are willing or can afford to pay for your weather station is a factor that will impact all of the factors that remain. In general, the more money you can invest in a personal weather station the better it is. However, comparing specifications is definitely worthwhile because the old adage "you get what you pay for" is not necessarily true in all cases.
- *Installation Issues.* The site where you intend to install your personal weather station needs to be evaluated, taking into consideration the distance from the indoor display console to where the sensors have to be located. For cabled weather stations, the length of the cable that comes with the unit will determine the maximum distance. Some manufacturers do offer extension cables for their cabled weather stations. Wireless home weather stations have a maximum 'unobstructed' or 'line of sight' range rating, which is diminished by the type and number of building materials the signal must penetrate. A rule-of-thumb for a typical installation is one half to one third the distance of the unobstructed range rating. Possible sources of interference should also be taken into consideration with the installation of wireless weather stations. The altitude where your weather station is going to be installed can also be an issue, and you should be aware that many weather stations are limited to operating at 6000 feet or below with regard to accurate barometer readings.
- *Accuracy, Resolution and Range.* These three factors are of primary importance when selecting the best weather stations. Accuracy is how close the displayed measurement reading is to the true measurement value (eg ±1°F for temperature). Resolution is the smallest increment that the unit is capable of measuring and displaying (eg 0.1°F for temperature). Range is the minimum and maximum limits the unit is capable of measuring within (eg -40°F to +150°F for temperature).
- *Update Interval.* This is the rate at which the personal weather station updates the display. Update intervals can vary significantly between units, from as often at once per second to as little as once every three minutes or even longer. Watching the wind data change every three minutes during a variably windy day, or likewise the rain data during a heavy rainfall, could limit the enjoyment you receive from your weather station.
- *Weather Forecasting.* If available, many commercial weather stations base their forecast on the barometric pressure trend (rising or falling), which is not as accurate as taking other variables into account. The more sophisticated professional weather stations takes into account not only barometric pressure, but also wind, rainfall, temperature humidity, and even the latitude and longitude of the station, resulting in a much more accurate forecast.
- *Historical Data.* Most basic weather stations and the less expensive complete weather stations display current data and very little in the way of historical data, perhaps the high and low readings for a period of time (often between manual resets). Weather stations that feature barometric pressure often include a graphic display of the trend for the past 24 hours. More extensive historical data retention is usually only available in a high-end professional weather station. For example, the Davis Vantage Pro2 series can display the highs and lows (and / or totals or averages) for nearly every weather variable for the last 24 hours, days, months, or years!
- *Computer Interface.* If you are interested in connecting your weather station to a computer you will most likely need to purchase a complete weather station, and not all of those come with that capability or option. With computer weather equipment (interface and software), you can record and graphically display weather variables at an interval that's typically user selectable. Depending on the software you can make forecasts, graph weather trends, post data on the Internet, or even send email alerts. Some manufacturers include a 'data logger', which has a built-in memory to store weather data for later retrieval. This allows you the flexibility of not having your computer on software running at all times.
- *Weather Station Review.* There is a lot of information available on the Internet about specific models of weather stations. We suggest you perform a search for some reviews on the models you're considering.

the weather will be like today. We are good storm spotters because we stay informed. We stay informed so the weather doesn't take us by surprise. We stay informed so we are ready to go when called on.

A valuable tool for the storm spotter is a home weather station. Today home weather stations come in all varieties from the simple barometer / thermometer that hangs on our wall in our home or out in the garden to weather stations that are digital and provide information on almost all facets of the current weather condition. If you have never purchased a home weather station or weather instruments for home use the current selection can be quite daunting. A commercial vendor of weather equipment has an excellent guide to choosing a weather station (see sidebar on the previous page).

Technician collecting weather data for the US Department of Agriculture. (Scott Bauer, US Department of Agriculture, photo)

REFERENCES

[1] NOAA Weather Radio (NWR): **www.nws.noaa.gov/nwr**
[2] Bob Bruninga, WB4APR: **www.aprs.org**
[3] Walter Fields, W4WCF, *GPS and Amateur Radio*, ARRL, 2007.
[4] Steve Ford, WB8IMY, *ARRL's VHF Digital Handbook*, ARRL, 2008.
[5] *EchoLink*: **www.EchoLink.org**
[6] Jonathan Taylor, K1RFD, *VoIP: Internet Linking for Radio Amateurs*, ARRL, 2nd edition, 2009.
[7] *Google Earth* Web site: **http://earth.google.com**
[8] T Wailgum, "Minneapolis Bridge Collapse: Why Cellular Service Goes Down During Disasters", CIO, August 3, 2007:
   www.cio.com/article/127901/Minneapolis_Bridge_Collapse_Why_Cellular_Service_Goes_Down_During_Disasters
[9] D Crowe, "Overload!" Cellular Networking Perspectives, Q4 2006:
   www.cnp-wireless.com/ArticleArchive/Wireless%20Telecom/2006Q3-Overload.html
[10] J Devonshire, M Flannagan, "Effects of Automotive Interior Lighting on Driver Vision", University of Michigan, Transportation Research Institute, March 2007.
[11] NWS Web site: **www.weather.gov**
[12] Storm Prediction Center (SPC) Web site: **www.spc.noaa.gov**
[13] AccuWeather: **www.accuweather.com**
[14] Figures given by AccuWeather
[15] WeatherBug: **http://weather.weatherbug.com**
[16] Stormpulse: **www.stormpulse.com**
[17] Weather Underground: **www.wunderground.com**
[18] The Weather Channel: **www.weather.com**
[19] Spotter Network: **www.spotternetwork.org**
[20] WeatherTAP: **www.weathertap.com**
[21] WXSpots download: **www.wxspots.com**
[22] GRLevelX downloads: **www.grlevelx.com**

Chapter 4

Training

Most storm spotters, whether or not an Amateur Radio operator, have taken a SKYWARN class. Some have gone farther with Advanced SKYWARN. A few have taken both, participated in exercises, are NIMS or ICS certified, etc. One area where Amateur Radio operators differ from others is with training. It is rare to find one who is *just* a storm spotter. Many hams active in storm spotting are also active in ARES or RACES, their local CERT team, or serve as volunteers with emergency management, Red Cross, or a public safety agency. Because of our involvement in public service and emergency response, training matters. As storm spotters, the more we can get, the better prepared we will be for an emergency.

In this section we thoroughly look at training for the Amateur Radio storm spotter, going beyond the basic level to better prepare ourselves. We divide training up into three basic areas: foundation, readiness, and continuing education. Much of this may overlap with training needed for other public service functions such as ARES, RACES, or CERT.

FOUNDATION TRAINING:

SKYWARN AND ARECC

Without a doubt, the first training any storm spotter should receive is through the NWS SKYWARN program. Repeating SKYWARN history from Chapter 1 is unnecessary, but it is worth looking at some current information about the program. As of 2010, NWS SKYWARN has trained around 280,000 spotters across the United States to provide information to NWS offices about local storms. This, combined with insight from radar and other resources, aids timely and accurate NWS watches and warnings.

Besides Amateur Radio operators many other groups have a vested interest in severe weather and need accurate and timely information about it. Groups such as police, fire fighters, emergency medical services, public safety dispatchers, media, school officials, hospitals, churches, nursing homes, and concerned citizens all make up the corps of SKYWARN volunteers.

Local NWS offices provide free SKYWARN training administered at each office be a Warning Coordination Meteorologist (WCM). The training usually takes only a couple of hours, covering, basic information about severe weather: thunderstorm development, identification and structure; identifying storm features; report methods and standards; and basic severe weather safety. The WCM also may customize training to meet local severe weather readiness needs, for example, nor'easters along the North Atlantic coastal area wouldn't relevant in Iowa.

Many NWS field offices also offer an advanced SKYWARN class that, like the basic course, can be structured to meet local storm spotter needs. It may take a few hours or spread over two or three days. The advanced course builds on from the basic one, providing more details on different types of severe weather, safety, what to report, radar interpretation, and spotting resources. The precise course curriculum is determined locally. Check with your NWS office to see if the advanced course is offered and what it includes. The NWS's

Storm Spotting & Amateur Radio

Advanced Spotter's Field Guide offers a general overview of the program.

In some areas there is a need for additional training because of how and to whom reports are sent. For example, spotters may report to net control, who then relays to the local emergency management agency and the National Weather Service. Spotters may also need training on inter-action of local agencies during severe weather, local activation procedures, and net protocols. Some in-house training, on top of the NWS SKYWARN program, can help your SKYWARN group effectively respond to severe weather.

THE ARRL EMERGENCY COMMUNICATIONS COURSE

In 2001 the ARRL began an online continuing education program for Amateur Radio operators in emergency communications. The Amateur Radio Emergency Communications Course (ARECC) was later broken into three different levels to cover the amount and diversity of information available about the subject. The Level 1 course, still available, covers a wide range of basic topics important to Amateur Radio emergency communications: served agencies, network theory, communication skills, message handling, nets, operations and logistics, etc. Levels 2 and 3 built upon the subjects and skills learned in Level 1.

Whether you are part of an ARES or RACES group or a SKYWARN storm spotter, this is essential training. As we discussed elsewhere SKYWARN storm spotters often double as ARES / RACES members.

In 2009, Levels 2 and 3 were merged, with changes to the curriculum that included courses from the Department of Homeland Security and the Federal Emergency Management Agency. The purposes of the redesign were to better prepare ARES leaders for the current emergency management environment and address future government credentialing issues that may come.

Credentialing is a serious issue in emergency management today. Any disaster, whether man-made or natural, involves a tremendous amount of responders. Many come from public safety, public works, and relief organizations, and from the local, state, federal, and tribal levels. Many involved in disaster response are volunteers, including Amateur Radio operators. During widespread events, such as hurricanes, many volunteers come from outside the affected area, sometimes from thousands of miles away. Organization and security in the disaster zone are critical. Regardless of the responder's origin, paid or volunteer, it is vital to know their identity, qualifications, role, and who they represent. Too often responders arrive with little or no verifiable credentials. While in Louisiana for hurricane Gustav I spent an entire week working in an emergency operations center without any credentials other than a badge with my name and call sign. Standardized credentialing seeks to prevent such situations.

READINESS TRAINING

Emergency Management Courses

A valuable training resource for the storm spotter and emergency communications is the Federal Emergency Management Agency (FEMA). FEMA offers free online training, courses ranging from basic level and citizen awareness to advanced responders. Many of these courses are great training resources for storm spotters.

Besides FEMA, many state and local emergency management agencies offer training classes. Examples include: Community Emergency Response Team, Weather for Emergency Management, NIMS / ICS, Hurricane Readiness for Inland and Coastal Communities, floods, and disaster recovery. In some areas you must be associated with a public safety or emergency management agency to take these types of courses. To find out course requirements and costs contact your local or state emergency management or homeland security office. Over the last decade or so ICS / NIMS increasingly has become mandatory training for many emergency responders.

In 1970 a wildfire in California raged for 13 days, killing 16 people, destroying 700 structures, and cost $18 million per day. Several entities involved in the fire response assembled to form a better and more efficient concept for fire response: the Incident Command System. Over time, the use of ICS spread to cover a wide range of incidents, including floods, earthquakes, hazardous materials, and aircraft crashes. For

FEMA Courses for the Storm Spotter

Course	Title
IS-7	A Citizen's Guide to Disaster Assistance
IS-22	Are You Ready? An In-Depth Guide to Citizen Preparedness
IS-100	Introduction to Incident Command System
IS-102	Deployment Basics for FEMA Response Partners
IS-120	An Introduction to Exercises
IS-200	ICS for Single Resources and Initial Action Incidents
IS-242	Effective Communication
IS-288	The Role of Voluntary Agencies in Emergency Management
IS-292	Disaster Basics
IS-317	Introduction to Community Emergency Response Teams
IS-324	Community Hurricane Preparedness
IS-700	National Incident Management System (NIMS), an Introduction
IS-800	National Response Framework, an Introduction

Storm Spotting & Amateur Radio

FEMA offers a number of emergency management courses.

several decades there have been some attempts at blending ICS and other incident management systems into a national standard. In 2003, President Bush issued the Homeland Security Presidential Directive 5 (HSPD-5) which ordered the development of the National Incident Management System (NIMS), taking ICS a step further. The purpose of NIMS is to provide "a consistent nationwide template to enable Federal, state, tribal, and local governments, non-governmental organizations (NGOs), and the private sector to work together to prevent, protect against, respond to, recover from, and mitigate the effects of incidents, regardless of cause, size, location, or complexity. This consistency provides the foundation for utilization of NIMS for all incidents, ranging from daily occurrences to incidents requiring a coordinated Federal response." NIMS and ICS training are available through FEMA online courses.

Community Emergency Response Team (CERT) was developed by the Los Angeles Fire Department in 1985, under the idea that during the early stages of a disaster, citizens may be on their own. If some in the general public had training in basic disaster survival and rescue skills they would be more likely to survive until responders could arrive. With the assistance of LAFD, FEMA's Emergency Management Institute expanded the program nationwide in 1994 to include all disasters. In 2003 President Bush initiated the Citizen Corps as a way for all American's to become involved in homeland security efforts. CERT was a perfect vehicle for this program.

CERT training covers basic disaster preparedness and response skills. Disaster related training includes; preparedness, fire suppression, medical operations, light search and rescue, psychology and team organization, and a simulation. Through the Citizen Corps Web site www.citizencorps.gov/cert you can find resources that include what is needed to start, train, and maintain a CERT.

CPR / First Aid / AED

Medical emergencies can arise whenever we are active as storm spotters. A storm spotter could encounter someone struck by lightning, a vehicle accident, injury due to hail or, in large scale disasters, mass injury situations. Storm spotting does not qualify us to render medical aid, but some basic, valuable training can prepare us to deal with medical emergencies until responders can arrive.

The Red Cross, American Heart Association, hospitals, or fire departments offer classes in most areas, including CPR and basic first aid.

CPR is a combination of chest compressions and artificial respirations into the patient to keep blood flowing through the heart and oxygen into the blood. It is not likely to restart the heart, but keeps oxygenated blood flowing to prevent tissue death and buy time until the heart can be restarted.[1].

Most basic first aid courses cover simple techniques that can be used to stabilize a patient until responders arrive, including: recognizing an emergency situation, caring for an unconscious victim, controlling external bleeding, sling and binder skills, removing contaminated gloves, and basic prevention of disease transmission. You should check with your local Red Cross or other basic first aid training provider for more details.

Automated External Defibrillators (AEDs), located in most public places, are used when a victim's heart has stopped. An AED administers a shock to aid in restarting the heart. Using a combination of lights, voice prompts, and text messages, it tells the person administering the AED when to shock. While it is not likely that a mobile storm spotter will be in a location where an AED is present, this is still useful knowledge. AED training often is combined with first aid and CPR training.

With luck you'll never need to use it, but CPR training is always a good idea.

CONTINUING EDUCATION

Keeping up to date and expanding our knowledge base helps us to stay effective and prepared. In this section, we will look at some of the abundant resources available to storm spotters in order to learn more about severe weather.

NWS Resources

A key role of the NWS is education.[2] Beyond SKYWARN training, the NWS offers a wide range of educational resources. Some are geared for children; others focus on safety and awareness; and some are quite useful to those interested in general meteorology and severe weather. Let's take a look at some NWS educational tools that a storm spotter may find helpful.

JetStream - Online School for Weather[3] is a Web site that offers material for educators, emergency managers, or anyone interested in weather and weather safety. The online curriculum covers the atmosphere, the ocean, synoptic meteorology, specific weather events such as tornadoes and hurricanes, radar, remote sensing, and weather on the Web. The online course features an introduction to each subject, an overview with visual aids, a frequently asked questions section, and review questions. You can go from one topic to the next in order or study which ever one interests you. The course is free and available to anyone.

The MetEd professional development series is a free online weather training from the COMET Program, which is part of the University Corporation for Atmospheric Research (UCAR). NOAA, the Air Force Weather Agency, Meteorological Services Canada, and the National Environmental Education Foundation sponsor MetEd, which is far more in depth than JetStream or SKYWARN classes. MetEd is intended for operational forecasters, the academic community, and "anyone interested in learning more deeply about meteorology and weather forecasting topics." Such topics include climatology, fire weather, radar meteorology, and space weather. You will need to sign up for a MetEd account to get started.

National Weather Service Safety Awareness Campaigns

Throughout the year the National Weather Service promotes weather safety through awareness week campaigns. These are done at the national, state and local level. Below are some examples:

National
- Flood Awareness Week
- Hurricane Preparedness Week
- Lightning Safety Awareness Week
- Air Quality Awareness Week

State and Local Level (can be a week or a day)
- Severe Weather Awareness
- Tsunami Awareness
- Winter Weather Safety Awareness
- Fall Severe Weather Awareness
- Thunderstorm Safety Awareness
- Wildfire Safety Awareness
- Tornado Drills
- Tornado Preparedness
- NOAA Weather Radio All Hazards Awareness
- Heat Awareness
- Summer Safety
- All Hazards Awareness

Check with your local NWS WFO for awareness weeks / days coming up in your area.

The NWS training division also offers some online educational resources covering a wide range of topics such as hydrology, AWIPS, and general meteorology. Most of this material is designed for practicing meteorologists; however, storm spotters and others interested in weather may find the topics useful and educational. The NWS Climate Prediction Center also provides educational resources on La Nina and El Nino, climate fact sheets and monographs, and monsoons.

NWS also offers a useful quarterly publication called *Aware*, designed to "enhance communication within the Agency and with the emergency management community". Current and past issues are available free through the NWS Web site, or you can sign up to have new editions emailed to you upon publication. Many local NWS offices also publish newsletters containing information of interest to storm spotters in their CWA. NWS regional offices are another source for news and information for storm spotters.

Beyond online thematic meteorology training there are other products available from the NWS that are useful for storm spotters, including a wide range of brochures on different weather topics. These brochures provide basic information as well as useful safety guidelines. Throughout the year, the NWS promotes awareness weeks for different

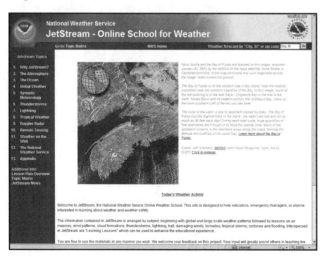

The JetStream - Online School for Weather Web site. (National Weather Service)

types of weather hazards. NWS educational resources are available through the main NWS Web site[2] and through your local NWS field office.

Colleges and Universities

Throughout the United States many colleges and universities have meteorology programs and research centers. While the primary function of these programs is for academic training and research, many have resources available to the public about meteorology and severe weather.

The University of Illinois' Weather World 2010 project (WW2010), is a free, online instructional guide covering meteorology, remote sensing, and reading and interpreting weather maps. The site also hosts open-ended projects, classroom activities and teacher's aids, each providing instruction on a wide range of related topics. For example, the meteorology guide contains sections named Light and Optics, Clouds and Precipitation, Forces and Winds, Air Masses and Fronts, Weather Forecasting, Severe Storms, Hurricanes, El Nino, and Hydrologic Cycle. Each topic contains text, images, video, and other dynamic presentations. The site is reviewed and edited by members of the Department of Atmospheric Sciences at the University of Illinois-Champagne-Urbana and the Illinois State Water Survey.

The WW22010 project is one of many educational outreach programs available through universities and colleges. Most schools with a meteorology or climatology program have material available through their Web sites covering different weather topics. Information on many of these programs is available through the University Corporation for Atmospheric Research and the American Meteorological Society Web sites.

Journals, Books and Media

Formal, scientific journals are another useful tool to expand your meteorology knowledge, keeping in mind though that most of their articles are written for an academic and professional audience. The publications will be rich in scientific, technical, and professional jargon; however, there can be a great amount of knowledge gleaned from these resources. You are not likely to find them at your local library or bookstore, however. Journal subscriptions are usually quite expensive, although some online journals are free, such as *The Electronic Journal of Severe Storms Meteorology*. Most college and university libraries subscribe to journals or electronic resources that can retrieve journal articles. However, most academic libraries require patrons to be faculty, staff, or students to access this information.

The American Meteorological Society is a source for scholarly material if you are not affiliated with a college or university. Founded in 1919, the AMS promotes "the development and dissemination of information and education on the atmospheric and related oceanic and hydrologic sciences and the advancement of their professional applications." Membership is available in several different classes; full, associate, or student. They also publish nine different weather-related journals in print and online and host conferences throughout the year, as wells as offer a wide range of products and services to their membership (currently about 14,000).

There are also several books that are of interest to storm spotters looking to expand their meteorological understanding. In 2009, the AMS published *The AMS Weatherbook; The Ultimate Guide to America's Weather*. Written by Jack Williams, a former editor of the *USA Today* weather page, the book covers a wide range of meteorological and climate information such as daily weather patterns, air pollution, and global warming.

The inquisitive storm spotter may also want to pick up a basic meteorology text book, like those used for freshman meteorology students. One that has been widely used is *Meteorology Today* by C David Ahrens. This textbook, which as of 2009 is in its 9th edition, is easily available through online sources such as Amazon. Another textbook and AMS publication, *Weather Studies: An Introduction to the Atmospheric Sciences*, by Joseph Moran, may also be useful to storm spotters. Prices for these textbooks are higher than for mass marketed books, but used copies abound.

There are also books dealing with specific weather events and historical weather events that are of interest to storm spotters. Author Peter Chaston has written several meteorology books. Chaston has written books on hurricanes, reading and interpreting weather maps, and thunderstorms, tornadoes, and hail. Chaston also has a book entitled *Weather Basics*. And we cannot forget historical weather events. Peter Felknor wrote about one of the deadliest tornado outbreaks in US history in *The Tri-State Tornado: The Story of America's Greatest Tornado Disaster*. This event, which impacted Missouri, Illinois, and Indiana, in March 1925, set the record for the deadliest tornado in US history and the longest tornado in US history (a 219 mile path). A single tornado, which measured an F5 with 200+ MPH winds, was responsible for 234 deaths, 2027 injuries, and the destruction of 15,000 homes.

There is a wide range of meteorology books available. Some are written for a general readership while others are written with meteorologists or meteorology students in mind. Most can be found in public or university libraries or online through sources such as Amazon. There are also many books dealing with weather history, both generally and specific weather events.

Magazines are another good source for meteorological reading. *Weatherwise* magazine is one such meteorology publication. Published quarterly, *Weatherwise* covers topics related to meteorology and climatology. They also publish photo contests, with photos submitted by readers, and an annual review of the previous year's weather. Another magazine source is *BAMS*, the *Bulletin of the American Meteorological Society*, the membership magazine of the AMS. Articles in *BAMS* are written primarily from a scientific perspective by members. Both Weatherwise and *BAMS* are often available through college and university libraries or through subscription or membership.

Another source of further reading is government reports on major natural disasters. Reports published by the US government, such as the Congressional report on the response to hurricane Katrina, are freely available online.

And for those who are interested in weather beyond the national level there are sources available through the World Meteorological Organization (WMO). The WMO, like the International Telecommunication Union (ITU), is an organization within the United Nations (UN). The WMO is "the UN system's authoritative voice on the state and behaviour of the Earth's atmosphere, its interaction with the oceans, the climate it produces and the resulting distribution of water resources." The WMO was founded in 1950 but traces its roots to the International Meteorological Organization which was founded in 1873.. Through the WMO Web site you can access publications relating to their primary areas of interest: meteorology, climate, oceanography, and water distribution.

Another source of training and preparedness material that cannot be overlooked is video resources. Companies such as The Weather Channel offer a wide range of videos on their Web site that are freely available. Topics covered include safety and preparedness, forecasts, and news. They also offer through online store videos of programs that air on The Weather Channel such as *Storm Stories* and *When Weather Changed History*. Of course these programs may also be viewed by tuning in to their scheduled broadcast. And National Geographic also offers a series of videos for sale that deal with weather: *Forces of Nature, Restless Earth* and *Tornado Intercept*. Also searching Internet sources can bring up a wealth of weather videos; some good, some bad. While on the subject of videos, don't forget Web sites of storm chasers. Many times they post online video clips from their chases.

Videos can be a great resource, but keep in mind that while some are quite instructive there are some out there that are not. Some may simply show severe weather as it is happening. Others may show people doing dangerous things. Always remember that safety is your number one priority. Unfortunately many of these amateur videos don't come with the warning "don't try this at home". When you're watching such videos you can easily turn them into training opportunities. You can assess the video on safety issues, reportable weather events, or look for specific weather phenomena.

SETS, EXERCISES, AND DRILLS

SKYWARN, like ARES, can benefit by conducting regular exercises and drills and taking part in the annual Simulated Emergency Test, or SET. Each year the ARRL Simulated Emergency Test provides an opportunity for Amateur Radio operators to test their emergency communications skills. Generally SET is planned for a weekend in the fall of each year, but some areas may plan to conduct a SET at a different time to meet local needs. For example along the Gulf Coast the fall is hurricane season so SETs are generally held in the spring. The planning of SET exercises is generally coordinated by section leadership; section manager, section emergency coordinator, district emergency coordinator, and local emergency coordinators. They also play a role in the review of test results.

The ARRL Emergency Communication Handbook[4] lists the purposes of the SET as to:

- Help amateurs gain experience in communicating, using standard procedures under simulated emergency conditions, and to experiment with some new concepts.
- Determine strong points, capabilities and limitations in providing emergency communications to improve the response to a real emergency.
- Provide a demonstration, to served agencies and the public through the news media, of the value of Amateur Radio, particularly in time of need.

And the goals of the SET are to:

- Strengthen VHF-to-HF links at the local level, ensuring the ARES and NTS work in concert.
- Encourage greater use of digital modes for handling high volume traffic and point-to-point welfare messages of the affected simulated-disaster area.
- Implement the Memoranda of Understanding between the ARRL, the users and cooperative agencies.
- Focus energies on ARES communications at the local level. Increase use and recognition of tactical communication on behalf of served agencies; using less amateur-to-amateur formal radiogram traffic.

A SET can be very beneficial to the SKYWARN storm spotter. A SET exercise may be based on a simulated weather emergency. SETs often focus on events such as tornadoes, hurricanes, and other natural disasters. The SET can also help local SKYWARN spotters to fine tune and improve their communications abilities for responding to severe weather. Spotters can learn to relay reports more accurately, net control and relay stations can identify ways to improve communications with the NWS. And it gives Amateur Radio operators a good chance to demonstrate their role during severe weather.

The SET also provides an ideal environment to work with served agencies such as the NWS and emergency management; this makes it possible for local amateurs and the served agencies to fine tune and develop their working relationship. Ultimately this can improve communications during a real SKYWARN activation.

The SET is also an ideal opportunity to try out and test new modes and equipment. We can try out local *APRS* networks, the *Echolink* connections with the local NWS office, the D-Star repeater system, or packet networks and test strengths and limitations. For mobile spotters and home-based spotters it provides a chance to test equipment and identify areas that need improvement.

The ARRL SET is not the only opportunity that we have to test our emergency readiness. At any time throughout the year local SKYWARN groups can organize a drill or exercise. This is a particularly good idea to do just before a known storm season approaches. Here are a few suggestions for planning and conducting a drill or exercise for your group:

- Coordinate with local emergency management. Many local emergency managers have experience in setting up drills and exercises and can be a great resource. And during an actual activation storm spotters likely will work with local emergency management.
- Coordinate with the local NWS office and their SKYWARN coordinator. If they have an Amateur Radio station in the office see if it can be manned during your drill so you can practice sending traffic to the office via their station.
- Try and have a drill or test shortly after a SKYWARN training session. Interest amongst SKYWARN spotters will be piqued by the training and you will likely have more involvement in the test.
- Make drills and tests a regular event. Too much time between SKYWARN activities can end up in lack of interest.
- Try something new with each test. If you haven't used *Echolink* or *APRS* or some other mode before try it in your next test and see how it works.
- Follow up each test with an after action report. Review what happened and make adjustments to plans as needed.
- Consult your local Emergency Coordinator, District Emergency Coordinator, Section Emergency Coordinator, and Section Manager for ideas. They may have experience in this area that can be helpful.
- When the drill is under way don't forget to announce clearly that it is a drill. Net control stations should make this announcement regularly.
- Should a real emergency situation arise during the drill stop the drill using an announcement that all participants clearly understand such as "real world" or "real emergency". Identify a key stop phrase before the drill and make sure all participants know what it is.
- Don't forget to test emergency power sources such as generators and battery backups. This should be done for home stations, net control, and repeaters.
- And after the drill make a final report for your SEC and ARRL. This is usually done by the local EC.

Drills and exercises are not limited to ARES / RACES only. Your SKYWARN group can use these as methods to train members, improve response and communications, PR opportunities, and ways to strengthen the tie between the NWS and the Amateur Radio storm spotting community.

Networking

Networking can be a valuable source of training and learning for the storm spotter. In many areas SKYWARN storm spotters are organized, through the local NWS office, into districts with district coordinators. Within each district there may be many local SKYWARN groups that activate during severe weather. Networking these groups together can be a tremendous asset. At the same time it can also be a tremendous challenge.

While it is not necessary to have a monthly area-wide SKYWARN meeting, it is a good idea to get area SKYWARN groups together perhaps on a quarterly or biannual basis. Topics for the meeting could cover a recap of activations, problems encountered during activation, assessment of communication between groups, talks on home weather stations, or arrange to have a NWS meteorologist there for a Q&A session. This meeting does not have to happen in person. Through technologies such as *Echolink, Skype*™, and wide area repeaters it is possible for SKYWARN groups to conduct a virtual meeting. If the local NWS office has these capabilities they can be networked into the meeting too.

Remember weather is dynamic; it can affect wide areas and involve dozens if not hundreds of storm spotters. We must also train to keep effective communication between SKYWARN groups going.

It is easy to think that training for storm spotters stops with SKYWARN training. SKYWARN training, basic and advanced, is the foundation for any storm spotter. These two courses are the 'must haves' for the serious storm spotter. However, training does not stop there. There are countless resources to help us learn more about severe weather, safety, emergency communications, and preparedness. Our training only stops when we refuse to learn any more.

REFERENCES

[1] "Learn CPR – You Can Do It!": **http://depts.washington.edu/learncpr/discr.html**
[2] For more information on the National Weather Service's role in promoting education refer to the NWS publication *National Oceanic and Atmospheric Administration Education Strategic Plan 2008-2009* available through the National Weather Service Web site: **www.weather.gov**
[3] JetStream - Online School for Weather: **www.srh.noaa.gov/jetstream**
[4] *The ARRL Emergency Communication Handbook*, ARRL, 2005 – 2007.

Chapter 5

Meteorology

So what does a storm spotter look for? What is a reportable event? What hazards might a spotter face? The United States faces no shortage of severe weather events. Year round severe weather is a threat. Typically spotters are called on to observe severe thunderstorms, tornadoes, hail, flooding, damaging wind, and winter weather. But in some areas spotters may also be called on to assist with hurricanes and tropical storms, coastal erosion, dust storms, volcanic ash fall, extreme heat, fog, or surf conditions. In this chapter we will look at common severe weather events that may require storm spotter activation. We will look at tornadoes and funnel clouds, thunderstorms, hail, damaging wind, floods, and winter weather. These weather events are common throughout most of the United States and are reportable by most spotter networks. We will also look at other weather events that may require spotters to be active. We will look at basic information about each weather phenomenon, safety issues for the spotter, and reporting criteria.

Some areas of this country are prone to types of severe weather that others are not, such as hurricanes along the Gulf Coast. What each spotter group is activated for will depend on what types of severe weather their areas are impacted by. In fact SKYWARN training keeps this in mind, allowing each local NWS office to tailor training to meet local needs. But remember: just because hurricanes hit coasts it doesn't mean other areas aren't affected. Hurricane Ike still packed hurricane force winds when it hit Southwest Ohio, over 700 miles inland! And while most deadly tornadoes impact the central United States, every state has experienced a tornado. In 2005 a tornado touched down near Sand Point, Alaska![1] While tornadoes in Alaska are rare, they do happen. Since 1950 only four tornadoes have been confirmed to have touched down in Alaska (three since 2004).

This chapter will provide the spotter with valuable information when observing severe weather. It will address some basic meteorological concepts behind the different weather events spotters may report on, reportable criteria and some real world accounts from spotters. Like the rest of this book, it is not designed to be a replacement for NWS SKYWARN training or other training provided by professional meteorologists, it is intended to complement the training that you receive.

THUNDERSTORMS

Thunderstorms are one of the most powerful and visible displays of natural energy. Many ancient cultures named gods after the fearsome power of lightning strikes. Even in our high tech age, lightning and other thunderstorm hazards are annually among the top natural disaster killers. SKYWARN storm spotters should have a good working knowledge of thunderstorm phenomena meteorology plus a high level of observational skills.

Before discussing the specifics of thunderstorms, let us investigate the processes of vertical motions in the atmosphere. This study will help one understand how many features of thunderstorms operate.

Suppose a parcel of air is forced to rise because it is crossing a mountain. The parcel will enter an environment of lower air pressure, and it cools due to expansion. If the volume of air is not fully saturated (no visible moisture present), it cools at 5.5° Fahrenheit per thousand feet, or 10° Celsius per kilometer. This is known as the *dry adiabatic lapse rate*.

If the parcel of air continues to rise, it will eventually reach *saturation*. This occurs when the parcel temperature and *dew point* become equal. At this stage the volume of air holds all the water vapor possible for a given temperature. Further cooling will reduce the parcel's ability to retain water vapor. The excess water vapor is condensed into liquid water, or perhaps ice crystals if the air is extremely cold.

When water vapor condenses it releases 600 calories of heat energy per cubic centimeter of liquid water formed. This effect is called the *release of the latent heat of condensation*. If the air parcel rises farther the released latent heat will partially offset cooling by expansion. The net result is that the vertical cooling rate is slower. This is known as the *moist adiabatic lapse rate*. The moist rate is not linear depending on many atmospheric variables. Sometimes a simple approximation of 3° Fahrenheit per thousand feet or 6° Celsius per kilometer is used to help visualize the process.

Now suppose the air parcel passes the mountain crest and starts to sink. It will be compressed by rising air pressures. The parcel temperature begins to exceed its dew point and no visible moisture will be present. The air will warm up at the dry adiabatic lapse rate as it heads to lower elevations.

Note this analysis assumes the air parcel is not interacting with the surrounding environment. Heavy precipitation occurs in many thunderstorms. Evaporation of the precipitation removes 600 calories of heat per cubic centimeter of water evaporated. This may cause an air parcel to become colder than otherwise expected. The importance of this will be discussed later.

Principles of buoyancy primarily determine the vertical extent of updrafts and downdrafts in convective clouds. If a rising air parcel is warmer than the surrounding atmosphere, it will rise like a hot air balloon on its own. If a parcel is not warmer than the surrounding atmosphere, it will be denser and tend to sink.

Meteorologists determine the existing vertical temperature, moisture, and wind distribution in the atmosphere by a *radiosonde*. This is an instrument package sent aloft by a weather balloon twice daily at hundreds of sites worldwide.

The observed vertical temperature pattern may or may not favor buoyancy of an air parcel. Suppose there is a layer of air in the atmosphere that is 80° Fahrenheit at its base. Now assume there is an air parcel of equal temperature at the base of this layer. Suppose 1000 feet higher the air temperature is 78°. Now let's force the air parcel to rise 1000 feet. It will cool by expansion to 74.5° Fahrenheit. Now remove the lifting force. What happens? The air parcel will sink because it is colder than its environment. This is an example of a stable air layer which is not favorable for buoyant updrafts.

There are rules for determining the stability of air layers. If the existing temperature decrease within the air layer is less than the moist adiabatic lapse rate, the layer is said to be stable. Note if the temperature actually rises with height, this is an inversion layer. Inversions are extremely stable. When they are close to the earth's surface, they trap pollutants near urban areas.

When the temperature difference from the base to the top of an air layer is between the moist and dry adiabatic lapse rates, it is said to be conditionally unstable. A rising volume of unsaturated air in this layer would not be buoyant. But if it is saturated it will arrive at the top of the air layer warmer than the environment. That means a saturated air parcel is buoyant in a conditionally unstable atmosphere.

If a layer of air cools vertically faster than the dry adiabatic lapse rate, the layer is unstable. Both saturated and unsaturated air parcels are buoyant in unstable environments. Unstable layers are usually shallow near strongly heated regions of the earth surface.

Now let's look at what happens in a typical vertically developed or convective cloud. One of a variety of reasons initiates upward motion of an air column. If the lifting process continues, the air reaches saturation forming a convective cloud base. This altitude is known as the *lifting condensation level* (*LCL*). The height of this level may be estimated from surface temperature and dew point data.

The LCL height in feet is approximately:
222 (Temperature minus Dew point in degrees Fahrenheit).

For metric units the LCL height in kilometers is:
1/8 (Temperature minus Dew point in degrees Celsius).

But will the cloud grow? It may grow somewhat further if there is a sufficient mechanical force to overcome negative buoyancy. But to get a really well developed convective cloud, at some point the existing temperature pattern must allow rising air columns to become buoyant. The altitude where buoyancy is first achieved is known as the *level of free (or unassisted) convection* or *LFC*. Upon passing the LFC air columns will rise like hot air balloons. The air columns will keep rising until they encounter an environment warmer than the column.

In most cases the lowest layer of the atmosphere, the troposphere, cools with increasing height up to 40,000 or 50,000 feet. Above this the stratosphere warms with increasing height, a form of inversion layer. Usually the top of the troposphere acts to restrain vertical cloud growth. The altitude where an air parcel or column runs out of buoyancy is called the equilibrium level. This is the theoretical convective cloud top, where a characteristic anvil-shaped cloud begins to spread downwind.

The actual altitude reached by buoyant air parcels depends on the range of altitudes where the parcel was buoyant plus the plus the degree of buoyancy attained. These two factors strongly correlate to updraft speed. If there is a powerful updraft the convective cloud may overshoot the equilibrium level by thousands of feet. The amount a thunderstorm overshoots the equilibrium level is one clue in estimating its intensity.

Note that in the rising air column analysis just completed, no interaction with the environment was considered. Suppose there *is* some interaction. One example might be encountering a *rain shaft*. Cooling by evaporation could occur, diminishing updraft buoyancy. The reverse situation applies in downdrafts. The cooling will accelerate them.

Modern meteorologists have many means of assessing atmospheric stability

Thunderstorm over New Mexico. (Craig Cornell photo)

and the potential for vertical motion. There are a wide variety of stability index products which are beyond the scope of this text. Interested spotters can learn more about them in many college level meteorology texts.

This discussion has strongly emphasized atmospheric stability. But what could change the existing or predicted stability? Any situation that increases the vertical temperature differences enhances instability. A very common source of unstable air is day time heating. Frequently the most unstable atmosphere occurs near or shortly following the daytime maximum temperature. A second instability source is cooling aloft. An approaching mid to upper level trough means colder temperatures at those levels.

A special note for anyone living in tropical ocean areas: there is very little day to night change in the surface temperature at sea. But the tops of existing clouds radiate more heat energy at night out to space. This means the tropical marine atmosphere is often most unstable late at night, resulting in the strongest convection during those hours. In contrast, surface cooling at night promotes atmospheric stability; an approaching mid to upper level ridge produces warming aloft, making the atmosphere stable.

Atmospheric stability conditions are a primary factor in determining thunderstorm potential. A second factor is the amount of moisture present. In extremely dry hot desert areas the atmosphere is often very unstable with turbulent strong updrafts and downdrafts. But there will be no clouds if the air is too dry.

Factors initiating updrafts (lifting mechanism) are important and have many varieties. Perhaps the easiest to understand is air forced to climb over rising terrain by the prevailing horizontal winds. Weather fronts divide air masses of different densities. A cold front vertically has a wedge shape which undercuts and lifts warm less dense air. Warm fronts have a gentle slope in which the less dense air glides over the colder air below. Ordinarily warm fronts are associated with layered and not convective clouds. The warm front discontinuity aloft is often an inversion layer. But if the warm air mass lifted is unstable and moist, gradual lifting may make air parcels buoyant. Embedded convection is possible with some warm fronts.

A third lifting mechanism, the dry line, often plays a key role in Texas and adjacent Southern Plains states. Suppose there is a general low level south to southwest wind flow during the warmer periods of the year. Air arriving from the Gulf of Mexico has a high water vapor content. Meanwhile air streams originating from northern Mexico will be hot but extremely dry. If the temperatures are similar, the dry air is denser and can act as a lifting mechanism and a focal point for new convection if conditions are otherwise favorable.

Small scale features are also important potential convection triggers. Near the Atlantic, Gulf of Mexico, and sometimes the Great Lakes, sea breezes may work their way inland on hot and humid days. These may trigger convection. Downdrafts from nearby thunderstorms act as mini cold fronts, and can set off new thunderstorm cells if the atmospheric conditions in the undisturbed air are otherwise favorable for convection.

Single-Cell Storm

Next we begin to describe and analyze the thunderstorm family. The simplest system is the *air mass* or *single-cell storm*. The sequence of events begins with a rising air column that forms a cumulus cloud.

If the atmosphere is relatively stable, the process terminates in a 'fair weather' cumulus cloud that is only a few thousand feet deep. However, if the atmosphere is conditionally unstable through a deeper layer, the updraft may achieve buoyancy causing the cloud to quickly grow taller into towering cumulus - see **Fig 5.1 (a)**. During this 'cumulus stage' the cloud contains primarily updrafts.

As the cloud builds, vast numbers of water droplets collide with each other, forming larger droplets. Before long they become rain drops producing both showers and downdrafts. Meanwhile the portion of the cloud in rain free areas keeps growing until nearing the equilibrium level, likely generating an anvil. Lightning and thunder are likely during this 'mature stage', as shown in **Fig 5.1 (b)**.

As the mature stage continues showers often become heavier and more widespread. Evaporational cooling near a vertical rain shaft main terminates buoyant updrafts nearby. The rain cooled ground makes the near surface layers more stable. Soon the updraft dies leaving only downdrafts in the 'dissipating stage'. The lower portions of the cumulonimbus cloud dissipate, leaving behind a remnant anvil for a while longer.

The processes described above for an air mass thunderstorm apply primarily when the winds aloft are not strong enough to horizontally displace storm updrafts and downdrafts. A significant percentage of thunderstorms fit into this category. Light mid to upper level winds are frequently found in the southern half of the United States during the peak of summer if no well defined weather front is near.

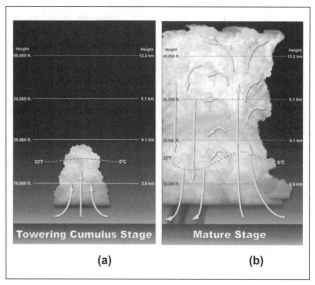

Fig 5.1: (a) Developing stage of a thunderstorm. (b) Thunderstorm mature stage. (National Weather Service)

Multi-Cell Storms & Squall Lines

As previously discussed, thunderstorm downdrafts may act as mini cold fronts. These may initiate new thunderstorm cells near ones that have recently dissipated. There may be a number of cells next to each other simultaneously at various stages of development. Storms acquiring this pattern are called *cluster* or *multi-cell storms*. This is the most commonly observed type of thunderstorm.

The third type of thunderstorm organization is the *squall line* which is an extensive line of multi-cell storms. Squall lines are solid or broken lines of thunderstorms that may be hundreds of miles long. They often propagate ahead of a strong cold front, but squall lines may form in other weather scenarios. The line of storms may align nearly parallel to the approaching front. In other situations the storm organization is more complex. Squall lines can interact with terrain features, individual strong surface gust fronts, plus changes in the overall stability and upper level winds. If the shape of an approaching squall line acquires a bow shape, portions of the line may harbor damaging surface winds.

All types of thunderstorms may theoretically become severe. The National Weather Service defines a severe thunderstorm as one that meets at least one of the following criteria:
- **1) Wind gusts of 50 knots (58 MPH) or higher;**
- **2) Hail 1 inch in diameter or greater;**
- **3) Any thunderstorm that produces a tornado.**

Air mass and multi-cell storms forming in low shear (horizontal or vertical changes in wind speed and direction) rarely become severe. The NWS estimates that only 1% of all thunderstorms attain severe status.

Squall lines, however, tap into extra energy sources that increase the possibility of severe weather. They may generate small to moderately large hail, extensive downbursts (downdrafts reaching the ground), and some weaker tornadoes. Squall lines need to be monitored very closely by NWS personnel and SKYWARN spotters.

Factors favoring squall lines are warm, moist, unstable air masses lifted by a strong quickly advancing cold front. In the upper levels an approaching trough will lower mid to upper level temperatures enhancing instability. There is often considerable vertical wind shear as the wind speed increases rapidly with height. In rapidly advancing squall lines the updrafts and downdrafts usually tilt with height. Heavy rain and hail often fall outside buoyant updraft areas. That permits updrafts to continually build the storm. Meanwhile the leading edge of the downbursts / gust front will keep lifting surface air in advance of the squall line. This can regenerate a squall line for many hours and sometimes for a day or two provided the atmosphere remains favorably moist and unstable.

Supercells

The most dangerous type of thunderstorm is the *supercell*. Such storms are the most likely ones to produce violent tornadoes, huge hail, and the most intense downbursts. Updrafts in the most extreme supercells can exceed 150 MPH. Quite a few thunderstorms are buoyant enough to reach over 50,000 feet. Other storm varieties such as the squall line may have a long duration due to tilting updrafts and downdrafts.

But in a supercell rotation is an additional factor that enhances storm intensity. The rotation can be generated from two mechanisms. In the lowest few thousand feet of the atmosphere a strong, usually southerly, wind increases rapidly and veers with height. This feature is commonly called a *low level jet*. The wind shear from this wind pattern tends to generate a spin around a horizontal axis. Now suppose this spinning air is lifted by a strong updraft. The axis of rotation becomes more vertical and imparts rotation into the lower portions of an evolving supercell thunderstorm.

Meanwhile aloft a strong jet stream six to eight miles above the ground will cross the lower altitude rising columns of air at sharp angles inducing more rotation. When everything comes together properly, these two features will cause an entire supercell thunderstorm to rotate. The rotating updraft, accompanied by relatively low atmospheric pressure, is known as a *mesocyclone*. Mesocyclones pose a significant risk for tornado formation.

Once a supercell is established, it can be self-sustaining for many hours. A single supercell may generate multiple varieties of severe weather along its track. Note that the precipitation rate associated with a supercell is not an accurate indicator of the amount of severe weather the storm produces. As long as most factors remain favorable, a supercell sometimes persists overnight despite the usual night time surface cooling. The general public tends to think that tornadoes are mostly a daytime hazard. Supercell produced late night tornadoes can be more deadly than normal if they catch people asleep and not prepared for them.

Thunderstorm Climatology

The annual frequency of all thunderstorm types varies according to season and location. A thunderstorm day is defined by the NWS as audible thunder heard.[2] In quiet rural environments thunder may be heard up to 25 miles away. Lightning can be seen up to 100 miles away at night from the tops of taller thunderstorms.

Annually most regions east of the Rockies experience 20 to 60 thunderstorm days per year. A majority of the storms occur during the spring, summer, and early fall, when the moisture and instability are greatest. A few Colorado and New Mexico eastern Rocky Mountain areas get up to 70 thunderstorm days per year. The north to south aligned mountain slopes are a significant source of lift when moist Gulf of Mexico air moves upslope.

From the Louisiana - Texas border to Georgia and points south, 70 or more annual days per year are also expected. The maximum frequency is just over 100 days per year in southwest Florida. The Southern State locations with high thunderstorm frequency have a long season of moist and unstable air. Sea breezes near the coast often initiate air mass thunderstorms in summer.

Florida is a unique case with sea breezes arriving from the Gulf of Mexico and Atlantic in the northern and central parts of the state. Farther south sea breezes may arrive from the Florida Straits. The convergence of sea breezes is a very effective lifting mechanism that explains much of Florida's high thunderstorm totals, very few of the thunderstorms started by sea breeze convergence become severe although sometimes they drop a great deal of rain. However, Florida can get severe thunderstorms associated with mid latitude strong cold fronts in the cooler months. Tropical cyclones in summer and fall may also generate thunderstorms meeting severe criteria.

Looking towards the West Coast, thunderstorm frequency is likely under-reported in sparsely populated mountain regions. The under reporting of the past is being corrected by modern lightning detection techniques. Continuing westward, a cold ocean current prevails along the West Coast. Normally the atmosphere is too stable for significant convection. But strong cold upper air lows can in some cases generate enough instability for coastal thunderstorms from late fall to early spring. A few summer thunderstorms can form over interior Southern California and Arizona when moisture flows northward from the Gulf of California.

Lightning

Outside of the mainland 'lower 48' states, summer lightning strikes often ignite huge forest fires in Alaska. Hawaii gets thunderstorms infrequently, primarily associated with winter storms. In contrast the Commonwealth of Puerto Rico is a thunderstorm hot bed. I personally observed an average of 160 thunderstorm days per year from 1994 to 2001 in Rincon, a northwest coast town. Puerto Rico has an east to west mountain range that is very favorable for lifting moist and unstable air. Typically 50 - 75% of the rainy season (May to October) days will bring a thunderstorm somewhere to western Puerto Rico.

Over land areas thunderstorms are most frequent from mid afternoon to early evening. This closely matches the times of the times of expected maximum heating and instability. Over tropical oceans thunderstorms are most likely late at night.

(Mike Corey, W5MPC, photo)

Lightning is a feature common to all types of thunderstorms. As a thunderstorm develops, varying parts of the cumulonimbus cloud acquire distinctly different electrical charges. Increasing updrafts and downdrafts bring opposite charged locations in proximity to each other. A lightning bolt discharges the electrical potential. Charge differentials between the cloud and the earth can lead to cloud to ground lightning. The overall intensity of a thunderstorm is often quite well correlated with observed lightning flash frequency.

Lightning strikes can be deadly, and are a hazard to storm spotters as well as the general public. Studies have shown an average lightning bolt carries 30 kilo amperes of electrical current, and values exceeding 300 kilo amperes have been observed; currents as small as 100 milliamperes can be fatal.[3]

Even relatively innocuous looking clouds can produce lightning. I will illustrate the point with a personal experience. When I was younger, I frequently went camping in North Carolina's Cape Hatteras National Seashore. One morning, around 6.00am, I awoke and noticed a towering cumulus cloud blowing in off the Gulf Stream. I wasn't concerned about the cloud, and I returned to my sleeping bag. Suddenly there was a blinding flash and a nearly instantaneous rifle shot crack of thunder. A solitary lightning bolt had struck a power line only 100 feet away. The cable sliced leaving a dangerous sparking live wire on the ground. There was no other thunder or lightning produced by this lone storm cell.

In addition to general personal safety, lightning is a special concern to amateur radio operators. Towers and external antennas must be properly grounded. Hams living where thunderstorms are common should consider well designed lightning protection systems. And don't forget the wisdom of unplugging sensitive electronic devices before thunderstorms approach.

Thunder is the result of super heating of air by a lightning strike. The heated air expands suddenly and then contracts, creating a resounding crash of noise. The time interval between a lightning flash and a peal of thunder can be used to estimate the distance to the lightning. Simply measure the number of seconds between the two events and divide by five. This gives you the approximate distance to the lightning in miles.

But don't rely on this method to keep you safe from the threat of a lightning bolt. You should be practicing lightning safety rules whenever threatening and turbulent clouds are observed. It does not need to be raining at your location for lightning to strike. Bolts of lightning can travel miles through rain-free atmosphere. Strikes have been observed up to 10 miles away from the parent cumulonimbus cloud.

Numerous additional problems accompany severe thunderstorms plus some that do not technically reach 'severe' criteria. Many thunderstorms of average intensity produce a short burst of heavy rainfall which is often beneficial. But on occasion a thunderstorm complex may move little for several hours. Or heavy precipitation cells may repeatedly cross the same region. This effect is known as '*training*'. An unusually moist atmosphere is an obvious ingredient favoring excessive rainfall rates. The presence of an orographic barrier or stalled weather front can cause storms to repeatedly develop in the same general location.

Heavy Rain and Flooding

The result of a heavy rain varies greatly depending on rainfall rates, duration of the event, and the nature of the surface the rain strikes. Dry and sandy soil can absorb a lot of water before there is a potential flooding issue. Moist clay soils that are already saturated will absorb far less water before flooding becomes possible. Highly urbanized areas absorb almost no water, and runoff problems can develop very quickly in high intensity rains. The usual local problem in flat regions is flooding of streets, culverts, and low lying homes.

When the terrain gets steeper, runoff moves rapidly downhill, concentrating into torrents. This inundation, called a *flash flood*, can be extremely dangerous and may develop suddenly. In many years flash floods are the number one cause of weather related fatalities.

In the arid western states hikers sometimes camp out in relatively flat but dry old creek beds. If there are thunderstorms over distant mountains, sudden walls of water may sweep the campers to their deaths. The worst place to get caught is in a steep narrow canyon. Flash floods can rise feet in seconds and there is no means of escape.

Mid latitude storms can occur with or without frozen surface precipitation. Some are heavy rainfall producers leading to general river flooding.

To conduct a flood potential analysis, one must consider an entire drainage basin from the point of interest to all locations upstream. Previous studies available will determine soil types and their absorption capabilities. It is possible to estimate the degree of soil saturation. Detailed mapping will show the varying elevations throughout the drainage basin. Real time water levels are measured by stream gauges.

When a significant rain event begins, hydrologists need to evaluate numerous factors. Critical information includes how much rain has fallen over a drainage basin, the rainfall duration, and the rain intensity. The analysis must cover scales ranging from the smallest tributary to the largest river. Real time rainfalls are measured by a combination of rain gauges and Doppler radar estimates. These combined with a dense network of surface observations, augmented by SKYWARN spotter reports, provide 'ground truth' that can overcome some Doppler radar limitations.

When a flood event occurs forecasters will evaluate a combination of real time data and numerical forecasting models. Appropriate warnings sent to local authorities permit timely evacuations and flood mitigation measures.

Flood forecasting in colder locations is especially challenging during winter thaws or the normal spring warming. The depth and water content of snow must be evaluated for an entire drainage basin. Very accurate temperature forecasts are needed to determine the snow melting rate. This rate of melting partly determines how quickly the released water reaches streams and rivers. The water moves most quickly over solidly frozen ground or other impervious surfaces. If the ground has thawed it may be able to absorb some water if the soil profile is not saturated.

Unexpected situations can alter flood forecasts. There is always a risk of a serious precipitation prediction error. Ice jams can act as dams creating unexpectedly high water upstream until the ice breaks up. Or man-made levees could fail. When a levee is breached, the flood covers a wider area, but the overall average water level is lower.

Hail

Hail accompanies quite a few stronger thunderstorms. Annually it is responsible for roughly $1 billion worth of property and crop damage. Even relatively small hail can damage or destroy crops. Larger hail stones dent motor vehicles, shatter glass, and cause personal injuries.

On occasion hail can get surprisingly large. Every year there are reports of baseball to softball size hail. The largest measured hail, *seven inches* in diameter, fell in Nebraska during 2003. This chunk of ice weighed almost a pound. A slightly smaller but denser hail stone from Kansas in 1970 tipped the scales at 1.7 pounds.

Hail typically falls in swaths or streaks that are considerably narrower than the source thunderstorm cell. Frequently hail or hail mixed with rain persists only a few minutes. Once in a while persistent hail can accumulate several inches deep, creating a winter-like landscape out of season. Heavy rain accompanying hail or nearby flash floods can push large hail accumulations around, creating locally much deeper hail measurements. High winds in some hail events can also create hail drifts.

Hail initially forms when an updraft lifts rain into below freezing portions of a cumulonimbus cloud. The chunk of ice can grow due to collisions with super cooled droplets. These are tiny water droplets in a cloud that may remain unfrozen at temperatures below 0° Celsius. Super cooled droplets freeze on contact with larger snow or ice masses in the storm cloud. In the case of hail these droplets produce *rime*, or cloudy-appearing ice.

Developing hail sometimes escapes the source downdraft and gets coated with liquid water where the temperature is above freezing. Then the hail may get recaptured by an updraft. The rain freezes on to the hail stone adding a layer of clear ice. Hail stones may cross the freezing altitude several times, growing as they transit updraft and downdraft regions. Mature hail stones frequently display alternating layers of clear and rime ice. The shape of large hail varies widely often displaying irregular protruding lumps or spikes.

At some point hail stones become too heavy to be supported by the existing updrafts, and they fall to the ground. Note that very large baseball size hail stones may fall at terminal velocities near 100 MPH. The fall speed can be faster if assisted by strong downdrafts. Large hail may also fall through strong horizontal winds. This causes the hail to strike exposed objects at an angle.

Hail at any one location is not a common event. In North America hail is most frequent in the Great Plains region, but it is possible in any state. The higher elevation Plains states are especially hail prone because the freezing level is often only a mile to a mile and one half above the ground. The hail stones have minimal opportunity to melt as they fall. Stations in warm humid climates near sea level have freezing levels at least three miles above the surface most of the time. The tropical coasts get comparatively infrequent hail (less than once per year) since most of it melts before reaching the ground. Many tropical regions rarely experience the thunderstorm types most likely to cause large hail.

Damaging Straight-Line Winds

Thunderstorm downdrafts frequently intersect the earth's surface and then spread outward. In milder thunderstorms the cool gust front provides welcome relief from a hot and humid day. In more intense thunderstorms the gusts may easily exceed the severe thunderstorm limit of 58 MPH. Every year there are documented reports of straight line winds well over 100 MPH.

The rush of air from a thunderstorm is known as a *downburst*. If the swath of damaging wind is 2.5 miles or wider, it is considered to be a *macroburst*, while swaths narrower than 2.5 miles are considered to be *microbursts* (**Fig 5.2**).

In most cases initial downbursts persist for only a few minutes. The most violent wind speeds occur just after the downburst intersects the surface. Wet downbursts associated with heavy rain often display a strongly sloping rain shaft close to the ground; while dry downbursts, accompanied by minimal or no rain, raise areas of blowing dust over open country.

The violent winds in downbursts can down trees, endanger mariners, and cause considerable property damage. Sometimes in squall line thunderstorms adjacent macrobursts merge causing lengthy damage swaths as the squall line advances. These events are sometimes called *derechos* (**Fig 5.3**). *Derecho* is a Spanish word for right. The name was given because many of the prolonged downburst events tend to deviate to the right of the course taken by their parent storms. Prolonged macroburst events are most frequent in the Great Plains region, but they are not too uncommon during the most thunderstorm prone months from the Midwest, to the Northeast, and Mid Atlantic regions.

A lot of people fly regularly, and they should be aware that microbursts are a special concern for pilots about to land. Normally on final approach a pilot flies a gradual descent on a predetermined glide slope angle. During descent the pilot gradually loses forward speed, planning to touch down near the end of a runway. The prime factor determining the lifting force of an airplane wing is the air speed relative to the aircraft motion. If a pilot encounters an approaching microburst head on, the wind pushing his airplane up (air speed) increases suddenly. The pilot finds himself well above the intended glide slope. An improperly trained pilot may take steps to lower his plane to get back on the desired glide slope. This action can be a fatal mistake. Suppose the pilot emerges on the other side of the microburst. A lot of things can go bad very quickly. The head wind will suddenly shift to a tail wind, and there will be much less lift on the wings. The plane will drop well below the safe glide slope, possibly leading to a crash unless corrective maneuvers are taken at once. These involve increasing power and attempting to escape the microburst with course changes. To make matters worse, the

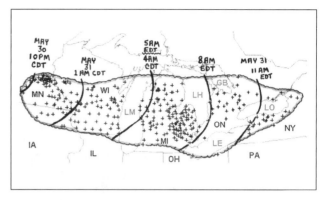

Fig 5.3: Progression of a *derecho*, May 1998. (National Weather Service)

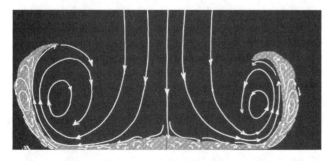

Fig 5.2: Diagram of a microburst. (National Weather Service)

Straight-line wind damage. (Bo Prince photo)

A Pilot's Take on Wind Shear

There are two types of wind shear: 'increasing performance' and 'decreasing performance'.

* **Increasing performance:** imagine a plane is sitting on the ground. The plane is not tied down and the wheels aren't chocked. Now a big wind kicks up. If a plane were light enough and the wind hit the wings straight on, the wind could cause the wings to create enough lift to make the airplane lift off the ground without the engine running.

Now put the plane in the air. The plane is flying along though stable air. Then, the plane flies into an area of weather and the wind is shearing in all different directions, like a frontal zone or a gust front. If the direction of the wind changes and causes a quick increase in wind on the nose of the plane the plane will also experience an increase in the airspeed of the plane. The result is an increase in lift and / or airspeed, but most likely both will increase at the same time.

* **Deceasing performance wind shear:** the anatomy of a downburst or microburst is air moving straight down and once it hits the ground it spreads out in every direction. Imagine water being poured into a basin. As you fly into the leading edge of the microburst you will always get increasing performance. In the middle the burst will cause an airplane to lose altitude because the air mass the plane is flying through is moving vertically downward (again imagine the water pouring out of a bucket). As the plane flies out the other side of the downburst the headwind has quickly become a tailwind. For example, a 50 knot head wind shearing to a 50 knot tailwind results in a loss of 100 knots of wind on the nose and therefore an equivalent loss of lift. This decreases the performance of the wing and can cause a plane to stall. This happened to Delta Flight 191 at Dallas - Fort Worth in 1985 and the plane was lost during approach.

In the event of a downburst, depending on where you are relative to the down / micro burst, you can experience one or the other or both. I experienced this flying into Amarillo (AMA) a few months ago. There was a cell on the far west side of the field and I was approaching from the southeast. With no warning my airspeed began increasing at an alarming rate. There were some visual clues, such as a cloud formation which would suggest a gust front. In addition to the increasing airspeed, the turbulence was moderate in intensity. My airplane has a wind speed and direction indicator. It quickly changed from a headwind of about 25 knots to a 35 – 40 knot crosswind from the left (so it changed from the NW to the SW) in just about one mile. It was probably my most exciting landing.

The other experience was on take-off from White Plaines, NY (HPN) on a winter's day. There was a very strong wind from the south. We were taking off to the east. The airport is surrounded by trees which work as a shield to the wind waiting above the trees. Once the airplane was airborne and flew above the trees, the turbulence began at moderate intensity. Then, the EGPWS (Enhanced Ground Proximity Warning System) which also handles wind shear alerts gave me a visual and aural warning. The next thing I heard was "WINDSHEAR, WINDSHEAR." The captain was the PNF (pilot not flying) and told me it was increasing performance, and to ride it out (we are trained to shout out our vertical speed trend and distance above the ground until we are clear of the wind shear. Because it was increasing performance we did not have to worry about the ground too much, we were being shot away from the ground at high speed. My initial level off altitude in my clearance was 2000 ft MSL. That is about 1600 ft AGL. Because of the turbulence and difficulty to maintain control, I was unable to control my altitude and exceeded my clearance limit by 1000 feet before I could reverse the climbing trend. In the event at Amarillo, I never got a wind shear alert, which surprised me given my airspeed increased about 40 knots in a few seconds.

I have never experienced decreasing performance wind shear, thankfully. I have trained for it, and have practiced wind shear penetration in the simulator. In the training exercise the airplane has good airspeed, but loses altitude because the air mass is moving vertically downward and not horizontal. In these cases, we are trained to select APR Power (Automatic Power Reserve), that is an aircraft specific term which spools the engine to maximum power, then we pitch the aircraft to the 'Pitch Limiter' and try to fly out of the wind shear straight ahead. Once out of the wind shear, the PLI (pitch limit indicator) will disappear and then the pilot has to bring airspeed and climb rate into control. There is no doubt that an Air Traffic Control clearance limit will be broken, but the Federal Aviation Regulations state that the pilot in command may deviate from the FARs to meet any emergency. Also, as I stated before, that while the PF (pilot flying) flies the plane through the wind shear, the PNF will shout out vertical speed trends, and height above ground from the radar altimeter.

pilot may have just seconds to react during violent turbulence and forces pushing downward on the airplane.

A microburst documented near Washington, DC almost involved Ronald Reagan when he was president. *Air Force One* had just landed. Six minutes later a microburst beyond the active runway caused the wind to speed up from 15 MPH to over 140 MPH in two minutes from one direction. Then briefly the wind speed dropped almost like in the eye of a hurricane. Next the winds jumped well over 100 MPH from the opposite direction. Finally they simmered down to normal speeds a few minutes later.

It is doubtful any airplane can survive such an extreme microburst if it catches the aircraft at the wrong moment. Fortunately such an encounter is much less likely now. Pilots are well trained on microburst flight techniques. Doppler radars can detect many potential downbursts, especially larger ones. Many thunderstorm prone busy airports have also installed low altitude wind shear alert systems (LLWAS) as a last defense to keep airplanes out of microbursts.

TORNADOES

Tornadoes are the 'top gun' in the severe thunderstorm arsenal. Despite their awesome power, research meteorologists still have a lot to learn about these fearsome phenomena.

The most dangerous tornadoes usually form in connection with supercell thunderstorms. The starting point is usually a mesocyclone, an area of low pressure and strong updrafts, best developed at mid levels. Falling pressures at the surface tend to cause the mesocyclone to expand downward. The descending mesocyclone is likely to contract, causing it to spin more rapidly. This potentially can lead to a funnel cloud and then a tornado. Meanwhile an adjacent downdraft, often known as a *rear flanking downdraft*, also descends to the ground.

Once the tornado touches down, it can intensify as long as there is an inflow of warm, moist air to support the parent mesocyclone. Eventually cool downdrafts wrap around the tornado, cutting off the supply of buoyant air. When this occurs, the tornado will weaken and dissipate.

Note that an intense supercell mesocyclone may rejuvenate multiple times if it finds new sources of moist and unstable air. As the parent thunderstorm advances it can be responsible for a series of tornadoes. The threat of additional tornado developments sometimes persists for hours until the supercell finally weakens.

Not all tornadoes form in connection with supercell storms. Sometimes converging low to mid level streams of air associated with other thunderstorm types acquire sufficient rotation to spin up a tornado. In squall lines tornadoes have a tendency to form on the southernmost cell or in cells just north of a break in the line. Tornadic rotations can also briefly occur in multicell storms.

The visual appearance of tornadoes and related clouds are well documented in NWS SKYWARN spotter training materials. You may wonder what causes a funnel to appear in

Tornado seven miles south of Anadarko, Oklahoma, May 1999. (NOAA library)

the first place. As you realize, a tornado is a microscale area of intense low pressure. Air that moves into lower air pressure cools by expansion. This expansion process is identical to the one in rising air parcels. If the cooling is sufficient, the air will cool to the dew point, and condensation will occur. In relatively dry air masses, there may be no easily visible funnel until the circulation contacts the surface raising dust and debris.

Tornadoes vary greatly in size, intensity, and path length. An average tornado is less than a quarter mile in diameter, lasts 10 minutes, and is on the ground for a few miles. A tornado tends to travel in tandem with the parent thunderstorm. In the United States a high percentage of twisters approach from the southwest or west. A typical 'Tornado Alley' type storm has a forward speed near 35 MPH, but the speed range can be from stationary to over 70 MPH. Tornado tracks are not always straight. They may swerve suddenly in response to changing air currents nearby. In rare cases tornadoes have even reversed course, such as the one near Dimmitt, Texas in June 2005 (below).

Tornado intensity is now measured by the *Enhanced Fujita Scale*. The range is from EF0, a gale force tornado, to EF5 which means three second gust speeds above 200 MPH. The tornado intensity is determined by NWS post storm survey. The estimated strength is based upon the damage produced accounting for the strength of construction materials.

The most violent three categories of tornadoes account for under 10% of those observed in the United States. The top category EF5 accounts for less than 1% of storms. A graphic example of EF5 effects was the near total destruction in Greensburg, Kansas in May 2007.

Nobody knows exactly the maximum possible tornado wind speeds or the minimum air pressures. However, a DOW (Doppler radar On Wheels) once measured a 302 MPH peak gust in a tornado near 100 feet above the surface in Oklahoma. An instrument package deployed in advance of another approaching tornado survived to record a 100 millibar pressure change in seconds as the twister passed.

The path length of a few large tornadoes may exceed 50 miles. The all time credible record was established by the 1925 Tri-State tornado. It was on the ground for 219 miles, as it spent four hours crossing portions of Missouri, Illinois, and Indiana. This tornado produced more documented fatalities (at least 695) than any other to date in our history.

Tornado shape, diameter, and general appearance can vary rapidly. At maturity some larger tornadoes are over a half mile wide. The record width based on storm damage surveys appears to be near 2.5 miles. Very wide tornadoes are sometimes difficult for the general public to recognize in the distance. They may appear to be huge turbulent masses of black cloud whose rotation may be obscured by vegetation, terrain, or man-made features. Large twisters are sometimes called *wedge tornadoes*.

In the majority of cases, there is a steady progression from a funnel cloud to a tornado ground contact. As the tornado strengthens it sometimes gets wider. During the weakening phases a dissipating funnel gets narrower and sometimes bends.

There does not appear to be a strong correlation between tornado diameter and peak wind speeds. A relatively narrow tornado can be extremely strong. Larger tornadoes often have several small vortices rotating inside the main funnel. Sometimes these are easily observed by storm spotters. The small vortices may account for some of the strange damage patterns seen in a tornado passage. Suppose there are two equally well built homes across the street from each other. After the storm one house may have relatively minor damage because it encountered 100 MPH winds. Across the street a mini vortex brought an additional 100 MPH making the total wind speed 200 MPH. The unfortunate second home is probably totally destroyed.

Tornado climatology is important information. All 50 states have tornadoes, but the relative frequency is highest in the 'Tornado Alley' states from Texas to the North Central Plains – see **Fig 5.4**. Extreme air mass contrasts are frequently present, setting in motion the factors that generate twisters.

A second high frequency area is Florida. The peninsula is a unique case. It is sometimes exposed to supercell thunderstorms in the cooler months of the year. Passing tropical cyclones may spin up tornadoes in their spiral squall bands.

Tornado south of Dimmitt, Texas, June 2005. (NOAA library)

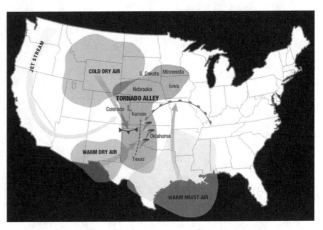

Fig 5.4: Map of 'Tornado Alley'. (National Weather Service)

The three way convergence of sea breezes may generate multicell thunderstorms capable of brief weak tornadoes.

In general most winter tornado formation is confined to the SE Atlantic region plus states bordering the Gulf of Mexico. Only these spots are likely to see occasional episodes of the required warm, moist, and unstable air. As spring approaches strengthening solar radiation more strongly heats the ground in the southern states. Meanwhile snow and very cold air retreats slowly in higher latitudes. Tornado frequency picks up over the Southern Plains and interior southeastern states. By April and May most of the United States east of the Rockies can see the recipe for tornadoes. Activity peaks in the Great Plains states, but violent tornadoes can form farther east. The April 1974 Xenia, Ohio tornado is a good example.

By summer the jet stream retreats northward, and there is often insufficient wind shear in Tornado Alley to produce the most violent storms. The northernmost states from the Dakotas to New England often have their greatest tornado frequencies during June, July, and sometimes August.

As fall approaches coastal states may get tornadoes produced by tropical storms and hurricanes making landfall. Otherwise tornado frequency tends to drop as the solar heating diminishes. However, in some years there may be a secondary tornado maximum in October and November. This is possible if surface storm tracks and jet stream positions are similar to those seen in spring.

Waterspouts observed over the ocean and sometimes lakes are a subspecies of the tornado family. There are two basic types. The 'fair weather' waterspout usually forms where surface waters are much warmer than the overlying atmosphere. This creates very unstable conditions near the ocean or lake surface. Now introduce a towering cumulus cloud and some wind shear near the surface. The shear may be produced by converging sea breezes or outflows from nearby decaying convective storms. As the air rises funnels may form in this environment. The typical fair weather waterspout is associated with towering cumulus or weak cumulonimbus clouds. Most of the time the water spouts are EF0 to EF1 intensity.

In certain circumstances supercell and squall line thunderstorms may move over lake or ocean waters. If these thunderstorms generate a waterspout it may be stronger than many fair weather waterspouts. The waterspouts originating from intense cumulonimbus clouds are called 'tornadic waterspouts'. Waterspouts are not counted when the NWS compiles their annual tornado statistics. However, if a waterspout of either type moves onshore, it is classified as a tornado and counted.

VALUABLE WEATHER TOOLS

SKYWARN storm spotters are called on to report on more weather conditions besides thunderstorms and their related events. Shortly we will look at some of these other weather phenomena and some things that fall more into the realm of geology and other earth sciences. Before we do so, we should discuss two of the most valuable tools to anyone interested in weather, weather radar and weather satellites.

Weather Radar
by Vic Morris, AH6WX

Modern Doppler weather radar provides many ways to evaluate weather systems of interest to SKYWARN spotters. Briefly I would like to review the history of weather radar. Radar as a military application was invented prior to World War II. Personnel on Oahu, Hawaii actually observed incoming aircraft from the Japanese Pearl Harbor raid in 1941 without realizing what they were seeing on their radar screen. After the war ended, the official position of the United States Weather Bureau (USWB) was to not mention the word tornado in forecasts because of the risk of causing panic (the USWB was the precursor to the National Weather Service). At the time, severe storm forecasting was in its infancy, and we did not have any means to detect tornadoes except by direct sightings.

The USWB recognized the value of radar in detecting precipitation, and the WSR-57 radar was deployed at many USWB offices during the late 1950s and 1960s. The early radars could only measure precipitation intensity, altitude, and range.

I had personal experience forecasting in the pre-radar period. Fast forward to a January night in 1971. I was a junior weather officer at the Navy Fleet Weather Central in Pearl Harbor. Ironically 30 years after the Pearl Harbor raid, there was still no weather radar in Hawaii. The only computer forecasting guidance available to Navy forecasters was the appropriately-named 'primitive equation model'. Forecasters did the best they could with surface observations, intermittent polar orbiting weather satellite coverage, plus some radiosonde reports. The closest one was from Lihue on Kauai which is WNW of the island of Oahu.

That particular January night I was on a 12-hour mid watch. In Hawaii most of the time we experience ENE trade winds with dew points in the 60s. However, at midnight there was a stiff SSW wind and a dew point over 70°. I knew from the surface analysis chart a strong low pressure area was

Waterspout off the Florida Keys, September 1969. (NOAA library)

passing a few hundred miles north of the Aloha State. The local air pressure was falling rapidly.

As part of my duties, I was responsible for preparing a local Hawaii forecast for US Navy and Marine facilities. As I was considering how to word my forecast, the latest Lihue sounding came in. Much to my surprise the report indicated extremely unstable air plus a vertical wind profile strongly supporting severe weather. Just one problem, at that time neither the Navy, Air Force, nor USWB forecasters had procedures for generating severe thunderstorms warnings.

So I went ahead an engineered a local forecast that called for torrential rains, frequent lightning and damaging shifting winds without actually saying the word "severe." Perhaps I exceeded my authority as a lowly Navy ensign. Luckily I was right or I could have been in serious trouble with my commanding officer. There were reports of 60 - 90 MPH wind gusts later on Oahu. After sunrise, a tornado injured a few people in a shopping center at Kailua-Kona on the Big Island of Hawaii.

The NWS embarked on a modernization program as computer technology developed and the old WSR-57 radars became obsolete. Now the WSR-88D Doppler weather radar systems and upgrades cover all 50 states, Puerto Rico and Guam. Deployment of the Doppler radar has greatly improved short term detection of many varieties of adverse weather conditions.

The Doppler radar emits very short pulses of UHF energy that produces reflections from precipitation sized drops, snow, ice, etc. The radar operator can adjust the sensitivity of the unit to meet his needs. The clear air mode is ultra sensitive and can detect unexpected items such as insects. Most events of interest to storm spotters are observed when the radar is in the precipitation mode.

The second objective for Doppler radar is to determine the motions of precipitation and local winds with respect to the radar. Doppler motion is detected by noting the frequency changes in reflected signals. To understand the Doppler principle, imagine an approaching whistling train. The sound appears to increase in pitch as both the sound waves and train are headed towards the observer. After the train passes, the sound is traveling in the opposite direction from the train. The observer hears a lower pitched whistle.

Modern Doppler radar systems can very accurately determine the velocities of particles directed towards and away from the site. Numerous applications of Doppler wind data are used for severe thunderstorm, hail, and tornado analysis. If a tropical cyclone is close to the radar site, the wind distributions can be accurately mapped out. There are additional Doppler applications for winter weather.

When the WSR-88D is operated, the radar dish may be elevated through 14 different angles from 0.5° upward to a maximum of 19.5°. The elevation choices permit the radar to probe clouds at various altitudes and distances from the radar site.

There are six NWS Doppler radar displays readily available to all users on the Internet. Other displays are available for those with special software, and you may see them from time to time on *The Weather Channel* and similar programs. I will cover the basic Doppler displays and their capabilities in this section. Later I will cover limitations that Doppler radar users should recognize.

Base reflectivity is perhaps the most commonly selected presentation. It shows the returns detected when the radar transmitter is elevated at a 0.5° angle. Composite reflectivity displays the highest intensity return on any of the 14 scans between 0.5 and 19.5° in the vertical. Short range displays of each reflectivity can reach up to 143 miles in favorable circumstances, and the long range reflectivity scan has a theoretical detection range of 286 miles.

The reflected returns are a color coded decibels (dBz) display. Blue and green are typically at the low end of the intensity scale. Moderate precipitation rates are typically yellow to orange. Heavy to extreme rates are typically red, purple, or white. Larger hail stones often display as extreme precipitation rates. Hail echoes often display very sharp, slightly irregular shapes. This contrasts to more gradual and variable shape changes in the display around heavy rain areas. Large hail will trip alarms in the local NWS Weather Forecast Offices.

A second set of displays shows the relative winds observed. A Doppler radar only can measure winds towards or away from the site along the 0.5° elevation of the radar beam. Green colors are typically a 'towards' site wind indication, and red are typically an 'away' measurement. To estimate the actual wind speed and direction intercepted by the radar beam one must know exactly where the radar site is located. The Doppler radar only shows full wind speed if the wind is blowing at right angles to the Doppler. As the wind direction deviates from the perpendicular, the presentation will shows winds lighter than reality. Some trigonometry and vector analysis can still provide good wind speed and direction estimates in the area of interest. If the wind direction is parallel to the Doppler, the operator will observe no velocity indication.

NWS Doppler radars can operate in the storm relative mode; this display compensates for storm motion. To illustrate, suppose you are on the bow of a ship moving 20 knots (nautical miles per hour) into calm air. You will feel 20 knots of wind. If there is a wind of 10 knots moving directly at you, the relative wind is now 30 knots. This is similar in concept to the microburst produces aircraft air speed changes described in an earlier section.

The storm relative Doppler display is extremely useful for locating the inflow and outflow regions near convective storms. Although the display usually lacks sufficient resolution to see tornadoes, it frequently picks out precursor mesocyclones.

A 'red flag' display is a couplet of small scale adjacent towards / away velocities embedded in an otherwise consistent velocity field. When the towards / away difference exceeds a preset value, alarms are triggered indicating a *tornado vortex signature* (*TVS*). If other indications suggest tornado potential, the local WFO will issue a tornado warning. Not every TVS results in a confirmed tornado, but forecasters are steadily improving their success rates and lead times. Recent policies now permit forecasters to issue warnings for only the portions of counties they feel to be at high risk. In the past most severe warnings had been issued for entire counties, creating an unacceptable false alarm rate for many residents, even if a small percentage of them did experience severe weather events.

The final two public Doppler radar displays are hourly rainfall and a forecaster selected storm total rainfall. While these displays are useful, they should be taken with a few grains of salt, especially if the estimate concerns long distance precipitation. Unfortunately Doppler radar has some significant limitations on its accuracy.

The first issue is generated by the altitudes of the radar beams (0.5 to 19.5°) and the curvature of the earth. If you live under 20 miles from the radar, the radar beams may not reach some local precipitation (the cone of silence), and the precipitation rate may be underestimated. You may get a more accurate picture from a somewhat more distant radar. In addition near the radar man-made structures or natural terrain may reflect some of the radar energy cone. This false return is called *ground clutter*. Computer algorithms can cancel out some ground clutter effects.

Typical flat terrain Doppler radars can detect the majority of precipitation within 80 miles and intense precipitation within 140 miles from your location. Sometimes at medium ranges radar will detect cloud level precipitation that evaporates before reaching the surface. As the distance to precipitation bearing clouds increases, the radar will intercept only the tallest portions of the clouds. The operator might see low intensity returns from cloud tops that are actually producing plenty of precipitations at altitudes under the radar beams. Satellite interpretation skill is a high priority where radar coverage is limited.

When a radar beam intersects very heavy precipitation, the energy beam is weakened before it can intercept more distant rain, snow, or hail. The more distant bad weather will show weaker than normal returns on the radar screen. This effect is called *precipitation attenuation*.

If you live near mountainous terrain, you are at the mercy of your local Doppler radar altitude. A significant range of horizontal directions or azimuths may be blocked off by higher ground. Ground clutter could obscure showers falling on the closest mountain slopes. Showers forming below the 0.5° upward radar beam may be undetected at considerably lower elevations. And you may not be able to see what is happening beyond the mountain crest either. But recall that the composite precipitation mode shows the strongest returns from any of the 14 radar dish elevation angles. That means the composite radar has a good chance of detecting heavier precipitation areas missed by the base level scan.

Anomalous propagation or super refraction occurs when strong inversion layers are present. The radar beam may be bent back to earth in unexpected locations. This causes false returns where no precipitation is occurring. On stable windy days some coastal radar get persistent returns from sea spray.

Perhaps worse, the bent radar beam may miss areas of real precipitation. If in doubt the best advice is to seek a second opinion. Refer to the latest surface observations and satellite imagery. As good SKYWARN spotters you are already carefully observing your local sky.

One last radar issue is worth mentioning for those exposed to winter weather issues. Heavier snow aloft will gradually melt into rain as it reaches elevations above freezing. The partially melted snow often gives a false indication of heavier precipitation known as a bright band. Checking local surface observations is the best means to determine the precipitation rates in these cases.

Weather Satellites

Weather satellite imagery has greatly enhanced our ability to analyze worldwide weather and make far more accurate predictions. SKYWARN spotters can gain a great deal of weather insight by learning some basics of satellite meteorology. The history of satellite meteorology has provided some interesting and exciting moments.

German scientists developed long range rocket capability near the end of World War II. During the subsequent Cold War, a space race developed between the Soviet Union and the United States. Space became a new frontier for military and civilian applications. The Russian satellite *Sputnik*, launched in 1957, was followed by the United States *Explorer I* satellite a year later.

Meteorologists recognized the potential of having an eye in the sky above the atmosphere. In 1959 the first weather satellite was launched, *Vanguard II*. It was designed to provide information on cloud cover and density in the atmosphere. Technical problems resulted in poor quality from the optical instrument on board. The *TIROS I* weather satellite launched in 1960 exceeded all expectations and became the basis for future weather satellites. Weather scientists quickly found out that many weather features are well organized and easily tracked from space.

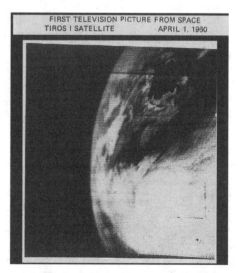

Image from TIROS I. (National Weather Service)

As the 1960s progressed, civilian polar orbiting weather satellites were launched. These provided relatively low resolution visible and infrared images. But the low frequency of satellite passes meant that weather features could go undetected for 12 hours or longer.

Meanwhile the military engaged in some ultra top secret programs to keep track of our Cold War enemies. Very high quality and high resolution images were generated by a secret satellite. This satellite played a key role in an incident involving *Apollo 11*. After our astronauts returned from their triumph on the moon, they were scheduled to splash down in a capsule in the tropical Pacific southeast of Hawaii. President Richard Nixon planned to be onboard the Hornet to greet our returning space heroes. Captain Hank Brandli, USAF, at Hickam Air Force Base had access to the very tightly compartmentalized ultra top secret imagery. He could clearly see that a tropical disturbance with 50,000 foot tall cumulonimbus clouds was heading into the *Apollo* recovery zone. If the astronauts parachuted into such weather, they risked death, and the United States risked a national disgrace. The young Air Force captain desperately needed to contact somebody to share his forecast information.

The Fleet Weather Central at Pearl Harbor was responsible for the Apollo recovery area forecasts. Fortunately the commanding officer Captain Willard S Houston, USN, had a security clearance that allowed him to see Brandli's data. Brandli appealed to Houston to do something. Captain Houston approached Rear Admiral Donald Dixon of the *Apollo* recovery group. The admiral did not have the clearance to see the secret images. And it would be a security violation for Captain Houston to disclose sensitive top secret information. But Captain Houston very persuasively passed on the warning to the admiral, despite a risk to his career.

Rear Admiral Dixon took up the matter with his Washington superiors. Fortunately the top military brass and senior NASA officials listened. The *Apollo* re-entry trajectory was altered, and the recovery task force was moved to a new safe location. A potential tragedy was avoided, and later Captain Houston was awarded a medal for doing the right thing.

The story above was declassified by President Clinton during the 1990s. The polar satellite data involved, the Defense Military Satellite Program (DMSP), is now readily available on the Internet.

Let's talk about modern weather satellite systems. As already noted, one common trajectory is a polar orbiter. The satellites typically take nearly 90 minutes to complete one orbit. The earth rotates under the satellites, causing the sensors to look at different locations on each orbit. That somewhat limits their ability to track weather unless a number of polar orbiters are available. On the plus side, polar orbiting satellites can follow trajectories just above the outer atmosphere, enabling them to obtain very detailed data. They are also a benefit to high latitude locations such as Alaska.

Geostationary satellites are fired into low latitude eastward orbits that match the rotation rate of planet earth above the equator. The satellite altitude is near 23,000 miles. At that altitude the satellite appears to hover above a fixed point on

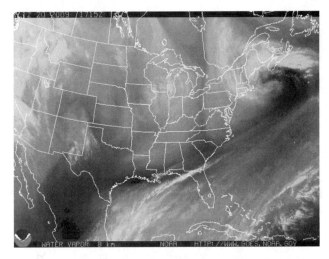

GOES water vapor image. (National Weather Service)

earth. The geostationary GOES and equivalent satellites can continuously monitor and track the weather below. The large distance from earth permits a GOES 'bird' to sense almost an entire hemisphere. But the data resolution is only moderately detailed. Given their consistent location above the equator, these satellites have less value to high latitude locations due to the earth's curvature.

Many varieties of data can be transmitted by both polar orbiting and geostationary satellites. Basically the satellites observe outgoing radiation at different wavelengths of interest to forecasters. Visible imagery is the easiest to understand. Imagine you are on a spacecraft observing the earth below. You see the tops of cloud patterns, but not what is underneath them. Strong convective clouds are normally easy to identify as they are often taller than other clouds present. The prime limitation is that one is unable to observe the weather where it is dark.

Infrared imagery senses the temperature of the cloud, water vapor, or ground surface radiating energy towards the weather satellite. The data may be processed in numerous ways, resulting in temperature coded color pattern displays. Some color schemes brightly display the coldest and presumably tallest cloud tops.

Others enhance only cloud temperatures falling in specific ranges. Shortwave infrared often depicts shallow to medium depth cloud patterns not well displayed by some conventional infrared displays. Water vapor radiates well on specific wavelengths, and modern imagery is good at detecting regions of moist and dry air at various levels.

The strong point for infrared imagery is 24-hour coverage. Rapidly cooling cloud tops are a sign of strong vertical motion and thunder storm development. Land and sea surface temperatures can be measured if no clouds are present. Such data is an important input for weather forecasting models.

Infrared does have one shortcoming. Suppose a cloud temperature is close to a surface temperature. In such cases it may be difficult to tell what the satellite is sensing. An example of this issue is detection of fog and other means need to be used to tell if fog is present.

There are some special weather satellite applications that deserve some attention. A modern generation of satellites contains microwave sensors. These are able to penetrate cloud canopies and detect high intensity rain cores. Microwave data can reveal hurricane eye wall structures, enabling more accurate hurricane intensity estimates. Other microwave sensors can estimate the surface winds at sea and approximate wave height.

OTHER WEATHER EVENTS

Storm spotters may be called on to observe and report a wide range of weather events and non-weather events that can be related to weather. Some of these events, such as dense fog, may occur just about anywhere. Other events such as volcanic ash fall are isolated to specific locales. Warning Coordination Meteorologists (WCMs) at NWS Weather Forecast Offices have considerable latitude in the types of information they may request from spotters. This allows the WFO to customize products to fit local needs. Spotters near the coast could be asked to report coastal flooding, beach erosion, and surf heights. In this section we will look at a few of these events and how they are important to the storm spotter.

Winter Weather

In this section we will discuss many of the impacts of non convective winter type storms. To better understand how such storms do their dirty work, it is important to briefly review air mass theories and the life cycles of middle latitude low pressure areas. These are often given technical terms such as the *wave cyclone* or *extratropical cyclone*. A cyclone is any type of low pressure area. An extratropical cyclone derives its energy from differing air mass densities. In contrast, *tropical cyclones* exist in a uniform air mass. They draw their energy from the release of vast amounts of latent and direct heat energy obtained from warm ocean waters.

The fundamental cause of *all* weather is unequal heating of our planet by the sun. Large expanses of air covering millions of square miles acquire similar temperature and humidity. These are air masses. Descriptive terms such as continental polar or maritime tropical give a snapshot of expected conditions.

When two contrasting air masses are next to each other, things get interesting. Cold dry air is dense, and it will attempt to undercut less dense warm humid air. In physics terms, there is potential energy awaiting release.

Suppose there is a stationary boundary of significant air mass contrast oriented east to west. And assume that the winds are blowing from 180° different directions on opposing sides of this discontinuity or weather front. What will release the potential energy?

The answer is gained from a vertical view in the troposphere. Suppose there is a region of divergence (air moving away from a location) in the upper atmosphere. The weight of the air column below lessens causing falling surface pressure. Air will rise to compensate for the air diverging at the top of the column. This is a recipe for bad weather. The falling pressure results in counterclockwise rotation of air around a point on the stationary front. Note if there is convergence aloft, the air sinks and surface pressure rises.

The wind speed normally increases with height in the middle to higher latitudes, although there are exceptions. The strongest winds are concentrated in a tube of air six to eight miles above sea level, called the *jet stream*. The jet stream is located above regions having a high horizontal temperature contrast. The polar jet stream is typically located over the middle latitudes, moving seasonally with the path of the sun. A secondary subtropical jet stream sometimes develops between 15° to 35° of latitude.

Meteorologists are especially interested in jet stream cores of maximum wind speed called jet *streaks*. Note that jet streams speeds can top 200 MPH. When a streak is viewed from above, the shape is often somewhat elliptical. The longest axis is aligned parallel to the wind maximum. Looking along the path of a jet stream the ellipse may be divided into quadrants in relation to the spot with peak wind speeds. Ahead and downwind of the jet maximum, the quadrants are labeled left front and right front. Behind the jet maximum the quadrants are labeled left rear and right rear.

Observations and theory demonstrate there is a consistent pattern of divergence and convergence associated with each quadrant. Divergence and upward motions are associated with the left front and right rear jet quadrants. If these pass over a stationary front, a well defined low pressure center may form. The right front and left rear quadrants are convergent. If these pass over the same stationary front, a significant low pressure area is not likely, but other forms of bad weather may still persist; however, if a new low pressure area develops, there will be increasing counterclockwise winds around it in the Northern Hemisphere. Frictional effects cause the wind to cross the isobars (lines connecting points of equal air pressure) slightly in the direction of lowest air pressure. As the wind picks up, the potential energy of the air mass contrast is converted into the kinetic energy of wind motion.

The surface fronts begin to move with respect to the low pressure center which tracks along with mid level steering currents. Advancing cold air undercuts warm air, creating a *cold front*. Warm air glides over retreating cold air as a *warm front*. The zone between the surface cold and warm fronts is called the warm sector.

Usually the cold front advances faster than the warm front, and the warm sector gets narrower. If conditions at all levels are favorable, the low will get stronger and wind speeds will increase. Eventually the cold front catches up with the warm front. This leads to an *occluded front*. In an occlusion surface cold air on opposing sides of the front meet. Remaining warm air is forced aloft. The air mass horizontal density difference greatly weakens. The low is usually near peak intensity when occlusion begins. As the occlusion process continues at increasing distances from the low, the entire circulation gradually spins down.

The mid latitude storm tracks and intensities follow

seasonal jet stream cycles. During the colder months the jet stream is usually stronger and farther south. Passing storms can affect almost any state depending on day to day weather pattern shifts. In the warmest months the jet stream is weaker and often not too far from the border between Canada and the United States. Mid latitude storms are most likely to affect the northern half of the country.

Precipitation

The first section of this chapter concerned primarily weather hazards generated by *convective clouds*. This section emphasizes problems produced by *layered clouds*. To really understand precipitation, one must look at microscale processes within clouds.

In the previous section it was assumed that condensation occurs when air is chilled to the dew point. But that is only part of the story. Moisture needs something to condense *on*, called a *condensation nucleus*. This is a very tiny particle of natural dust, sea salt, or man-made pollutant. Some nuclei such as salt actually *attract* moisture and are called *hygroscopic*.

The average cloud droplet is 20 microns in diameter, although there are significant size variations (a micron is one millionth of a meter). Then what causes precipitation?

The simplest case to analyze is a cloud entirely warmer than freezing (32° Fahrenheit or 0° Celsius). Both horizontal and vertical winds within the cloud cause the varying sized droplets to collide. Sometimes the droplets bounce off each other. In other cases they merge or coalesce. Cloud physics research shows that droplets containing opposing electrical charges are most likely to merge. As the droplets get bigger and heavier, they fall faster and merge more frequently with their neighbors. Showers will soon reach the surface. In some cases rain can fall only 15 to 20 minutes after a warm cumulus cloud forms.

All year in the tropics, plus sometimes during the middle latitude summer, freezing levels are near three miles above sea level. Rain that falls from clouds entirely warmer than freezing is referred to as 'warm rain'. Studies have shown that cumulus clouds as shallow as a few thousand feet deep can produce showers, especially if they contain salty hygroscopic nuclei.

Many clouds contain regions that are both warmer and colder than freezing, and numerous winter clouds are entirely below freezing. Such clouds may contain different varieties of water droplets and ice crystals depending on the vertical temperature distribution. In cloud regions warmer than 0° Celsius, the expected conventional water droplets form. But in clouds below 0° Celsius there is an unusual form of liquid water called *super cooled droplets*. Most people think of the freezing point of water as 0° Celsius. That is entirely correct for large volumes of water such as a lake surface or even ice cubes in a glass. But at the diameter of tiny cloud droplets, the water can remain as a liquid down to temperatures well below 0° Celsius. If one inventories a cloud at temperatures below 0° Celsius, there will be a mix of super cooled droplets plus some ice crystals. The fraction of ice crystals increases as one gets closer to -40° Celsius. Most of the naturally occurring freezing nuclei 'activate' at about -10° C at which point the cloud consists entirely of ice crystals.

When falling ice crystals contact super cooled droplets, the water freezes immediately. The ice crystals grow at the expense of moisture in the previous super cooled droplets. The growth of a falling ice crystal is known as *accretion*. Once sufficient growth occurs, the complex of ice crystals becomes a snowflake.

As snowflakes enter regions above freezing beneath the cloud they will melt into a cold rain. The process of rain formation that began with ice crystals is known as the *Bergeron process*.

Snowfall

The majority of snowfall occurs in connection with passing mid latitude storms. Quite often during winter the warm sectors of such storms will be warmer than freezing. But as the low pressure area moves along its path, warm air will be lifted gradually ahead of the surface warm front. Snow is likely from east through northwest of the developing low center if the entire region is colder than freezing. If only some portions of a winter type storm are colder than freezing, the situation can get complicated. I will illustrate the points with some examples from New England, where I grew up. The examples provided here can be applied elsewhere in similar cases.

The first hypothetical storm directly approaches Cape Cod, a peninsula off southeastern Massachusetts, from the southwest. The open ocean near Cape Cod is usually 35 to 40° Fahrenheit during the winter. If the wind is from the east, any early snow on Cape Cod will quickly turn to rain. Meanwhile away from the coast in interior Southern New England there will be a swath of heavy snow. Farther northwest a light powdery snow tapers off well away from the storm center.

In scenario #2 the low center moves across central Massachusetts. Locations east of the low will see primarily rain. West and north of the storm track expect mostly snow, although other forms of frozen precipitation are possible. I will cover those later. In many cases the heaviest snow occurs

A blizzard in the Evergreen, Colorado area with near white out conditions. (National Weather Service)

50 - 100 miles northwest of the low pressure center track. Once the low center moves by, the winds turn northwest, and it gets much colder. Areas of slushy wet snow will freeze creating lots of slippery ice. Snow squalls are possible in areas that received mostly rain. I used to call mixed precipitation followed by a freeze a 'slop' storm. There are likely unprintable terms for the same mess!

In scenario #3 the low center passes 50 - 100 miles east of Cape Cod. If the air near the Cape is cold enough, it will be their turn to get a heavy snow. Locations farther inland will receive smaller totals.

Certain locations near water bodies such as the Great Lakes are susceptible to a special type of snow storm. The lakes become relatively mild in summer, and are slow to give up their heat as the weather gets colder. Suppose a very cold arctic air mass crosses an unfrozen large lake. The air is heated from below, and a layer of very unstable air develops. Lines of cumulus clouds build, and these may become very prolific snow makers. Quite often persistent low level wind directions cause lines of heavy snow bearing clouds to train over the same location for many hours. There is often a sharp divide between minimal to excessive lake effect snowfalls. Prolonged lake effect events can pile up to several feet - just ask any resident of Buffalo or Cleveland.

Snow storms are possible in portions of all 50 states. As expected the northern United States cities usually get the largest annual total snow falls for heavily populated areas. But snow flurries have been seen as far south as Miami. In Hawaii the taller peaks of Mauna Loa and Mauna Kea on the Big Island plus Haleakala on Maui get snow during most winters.

Freezing rain and ice pellets (*sleet*) deserve special attention. Suppose there is a slow moving warm front or one that has turned stationary. At higher altitudes the air glides up the frontal slope. Suppose some of the rising air is warmer than freezing. At your surface location it is below freezing. Rain develops initially in the warm air aloft. But if it falls through a considerable depth of below freezing air, it will freeze creating small relatively harmless chunks of ice called ice pellets.

If the near surface below freezing layer is shallow, the rain will not have a chance to freeze in the air. Instead it freezes on contact with all surfaces colder than 32° Fahrenheit. The result is very dangerous freezing rain. It quickly covers streets and encases power lines and tree limbs. Prolonged freezing rain events can easily produce over an inch of ice. That is a recipe for lengthy power failures, downed branches, and extremely hazardous driving conditions.

Given the proper temperature patterns, certain weather or geographical situations favor persistent freezing rain. A nearly stationary front may be aligned across a portion of the eastern or central United States. If the jet stream position is not favorable, no major storm center will form. Instead a series of weak lows may move along the front causing moderate lift. Surface locations on the colder side of the stalled front may get many hours of freezing rain.

Mountain ranges such as the Appalachians often act to dam cold air masses ahead of approaching warm fronts. The warm air may not be able to dislodge cold dense air in mountain valleys. These are locations very susceptible to freezing rain.

Let's look at winter weather hazards on a broader scale. The obvious problems of a heavy snowfall include difficult driving, closed schools and businesses, and prolonged flight delays. Matters get far worse if there is high wind.

The NWS defines a blizzard as snow or blowing snow accompanied by frequent 35 MPH or stronger gusts which lower visibilities to 1/4 mile or less. Motorists caught in a blizzard can be stranded for many hours. Drivers in blizzard prone areas should have winter weather survival gear in their vehicles. In blizzards snow drifts many feet tall form, and high winds can redeposit snow almost as quickly as plows can remove it. Hikers and others outdoors in open country can get disoriented and lost. Prolonged exposure may lead to frostbite and possibly death.

The high winds associated with a blizzard are produced by strong pressure gradients near the parent low pressure center. If a strong Arctic high pressure area moves in behind the departing low, blowing and drifting snows are likely well after the snowfall ends. During this post storm period rapidly falling temperatures are common. The mix of high wind and cold air creates dangerously low wind chill readings. Most residents of cold climates know they should dress in layered winter clothing, and limit their exposure to very low wind chill temperatures.

Nor'easters

Intense coastal mid latitude storms create additional hazards. Along the United States Atlantic coast, a fairly common cool season storm track is from near Florida to the Carolinas and New England. These are often called '*northeasters*' by Atlantic coast residents. Many of the storms reach gale to storm force (39 - 73 MPH), and a few may exceed hurricane force (74 MPH or greater). Very intense storms also frequently approach Alaska, Washington, Oregon, and Northern California during the same months.

Usually - but not always - the highest winds are at sea. If the storm moves rapidly, the main issue near the coast is a few hours of wind strong enough to cause minor damage. However, very high winds at sea will create waves large enough to endanger commercial fishing boats.

On infrequent occasions a very intense mid latitude storm can stall or move slowly on erratic path for days. The classic example of this is the 'Perfect Storm' of 1991. In the closing days of October that year, an unusually strong Arctic high pressure area set up over northern Quebec. It sent blasts of very cold air over the still relatively warm Gulf Stream south and east of New England and Nova Scotia. An extremely potent upper level trough created severe divergence aloft along a front southeast of Sable Island, Nova Scotia. Surface pressures fell very rapidly around an explosively deepening storm. Additional energy was also provided by the remnants of a late season hurricane moving into the area.

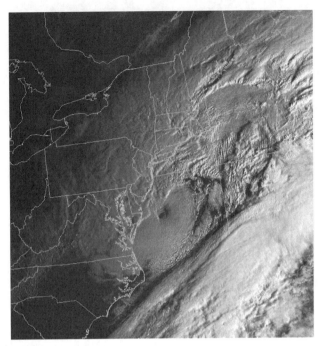

GOES image of 2006 Nor'easter. (National Oceanic and Atmospheric Administration)

By October 29 a well defined 970 millibar storm was established south of Nova Scotia. The air pressure was 75 millibars higher in Quebec. The great pressure difference created a very large region of storm to hurricane force winds over the northwest Atlantic. Usually middle latitude storm systems move towards the northeast or east. But the extremely strong upper trough near the storm caused the system to make a slow counterclockwise loop over several days.

The maximum wave heights observed at sea are explained by three factors. These include wind speed, duration of high wind, and the *fetch length* (a fetch is a region having relatively constant wind speed and direction).

During the 'Perfect Storm' there were unusually strong winds blowing for days over hundreds of miles of ocean. Extraordinarily high seas developed. A Canadian buoy south of Nova Scotia recorded a peak wave 101 feet high. This is the largest one ever measured by a weather buoy. The exceptionally high seas caused the fishing boat *Andrea Gail* to sink. Several other boats were also lost or severely damaged.

The winds were near hurricane force in gusts along the eastern New England coast. Storm tides (the sum of the astronomical tide plus storm surge) were several feet above normal. They may have reached 10 feet above normal on parts of outer Cape Cod. At Truro on the Cape the huge seas broke through a 100-yard wide 20 ft high dune to create a new ocean inlet. The inlet remained open for several months. Numerous New England shoreline properties were battered by waves and flooded.

Very large swells created huge surf in areas far removed from the storm. Almost unprecedented 20 foot breakers reached Palm Beach, Florida. Some 35 foot waves were seen off northwestern Puerto Rico.

The Pacific coast is exposed to large waves more frequently than the Atlantic coast. Winter storm tracks quite commonly aim potentially dangerous fetches at Hawaii, California, Oregon, Washington, and Alaska. Battering waves and the run up of ocean water periodically threaten coastal properties. Very large waves produce deadly rip currents that can pull swimmers out to sea. Never walk along a beach backed by steep wet cliffs during high surf. That is a warning the waves have reached a high elevation, and you could be caught by a breaker with no place to escape.

Beach erosion is a related issue impacting coastal communities. A prolonged period of heavy surf will lower the beach profile depositing the removed sand well offshore. As the beach gets lower, waves can attack the dunes behind them, perhaps undermining some structures. The worst beach erosion often occurs when a persistent gale is blowing across the coast at an acute angle for several consecutive high tide cycles. During one storm in the 1980s I personally observed a 100 ft width of beach disappear at New Smyrna Beach on the north central Florida Atlantic coast. This phenomenon will likely only be exacerbated in the coming decades due to the observed sea level rise attributed to global warming. After the storm some lucky residents found treasure coins and other artifacts from old Spanish shipwrecks.

Dense Fog

Fog is a weather problem that affects many people, and it can be highly localized. On occasion freeway drivers traveling at 70 MPH suddenly enter pockets of near zero visibility. If they fail to slow to a safe speed, multi vehicle pileups are very possible. Spotter reports can help forecasters better pinpoint locations for dense fog advisories.

There are several major fog types. Suppose at sunset the sky is clear but it is humid with little wind. Heat energy radiating from the surface will eventually cool the air to its dew point temperature. The result is *radiation fog*. It tends to pool in low spots, and it is often most widespread late at night through the early morning hours.

A second variety of fog is common near coasts with cold water. During the warmer months humid air blowing over the chilly oceans will get cooled to its dew point. This process creates *advection fog*. Advection means the horizontal motion of some atmospheric property. Both the New England coast and the Pacific coast in North America have high advection fog frequencies. Advection fog tends to thin out or dissipate by day over land. But it will likely persist just offshore. After sunset the coast cools off and the fog will likely return.

If humid air climbs a mountain range, the expanding air may cool to its dew point. *Upslope fog* is the result. Note an observer at a lower elevation will be looking up towards a cloud base.

During winter very cold and dry air can cross unfrozen waters. Moisture evaporates from the lake or ocean raising the cold air dew point; when the dew point and temperature become equal, fog forms. The technical name for the result is *evaporation fog*. But frequently this fog variety is referred

Early morning radiation fog. (National Weather Service)

to as '*Arctic sea smoke*', a very accurate description of its appearance.

NON-WEATHER EVENTS

Landslides

Steep slopes are prone to problems related to both weather and geology. Suppose there is a prolonged period of heavy rain falling on a deep layer of poorly consolidated soil. If the soil gets saturated, portions of a hillside may give away suddenly, causing a mudslide or a landslide. Areas devoid of natural vegetation due to recent wild fires or poor grading practices are particularly prone to sudden collapse. In steep rocky areas heavy rain can cause rock slides. These also can cause property damage and are a safety hazard.

Avalanches

Very deep accumulations of snow are common in mountain regions. If the snow overhangs a steep slope, it can collapse suddenly, starting an avalanche. Avalanches become very likely when snow gains weight from the seepage of nearby melting. There is almost no advance warning for a specific avalanche. Skiers and snow mobile drivers may be threatened. Ski patrols sometimes deliberately set off avalanches in controlled safe environments to reduce the risk.

Wildfires

Another significant hazard is fire weather. Every year numerous very large wildfires occur in our country. A long previous period of dry weather precedes most major fires. This situation is a normal part of some climate types. For example, Southern California normally gets very little rain from mid spring to mid fall.

The fire hazard greatly increases on dry windy days. Red flag warnings are issued by local WFOs when wind gusts are high and the relative humidity is low. The fire threat will be especially high if there are any lightning strikes into dry vegetation..

Many large scale weather scenarios can create conditions favorable for wildfires. One of the most notorious is the Southern California Santa Ana wind. The basic setup is a strong high pressure area over the Great Basin near the fall season; air blowing clockwise from the high pressure area heads towards the California coast. As the air sinks it warms by compression and the relative humidity sometimes drops to single digit values. To make things worse the winds can accelerate near mountains and canyons. Sometimes the gusts top hurricane force. Out of control blazes can spread very rapidly leaving very little time to evacuate. The risk to human lives has increased in recent years as more people build homes in high risk wildfire zones.

Volcanic Activity

Weather Forecast Offices in the Pacific coast states sometimes need to deal directly or indirectly with geological phenomena. These locations lie along the Pacific 'Ring of Fire', and volcanic eruptions are possible. Volcanoes tend to fall into two varieties. Most from California to Alaska tend to be moderately to highly explosive due to gas trapped under pressure in subsurface molten rock known as magma. The Mount St Helens eruption of 1980 serves as an outstanding example. Such eruptions not only produce lava, but they may also generate pyroclastic flows - superheated dense masses of volcanic ash and poisonous gases which can move down the slopes at speeds well over 100 MPH.

Volcanoes covered by heavy snow and glacial ice present additional hazards. If lava or a pyroclastic flow reaches deep ice, sudden melting will unleash tremendous flooding. Things get even worse if a crater lake breaches adding to the torrents. The sudden deluges contain volcanic ash, mud, and debris. These events are known as *lahars*.

Mount Rainier east of Seattle is a prime candidate for future lahars. Several hundred years ago geological evidence indicates that lahars dozens of feet deep reached some current cities of southwestern Washington State. Repeat performances are likely in the future from Mt Rainier, and perhaps other glaciated Pacific coast volcanoes.

Volcanic ash can spread considerable distances based on vertical wind profiles. Falling ash is a serious health hazard that impairs breathing function. Heavy ash falls can lead to roof failures and other property damage. The problem of roof collapse is exacerbated if the volcanic ash becomes wet from rain as it takes on the consistency of concrete. Suspended volcanic ash is a serious danger to aviation. The ash can be ingested by jet turbine engines. Engines can abruptly cut out and shut down, resulting in power loss and rapid loss of altitude. If a pilot cannot find reasonably clear air to restart the engines, a crash is likely. Accurate aviation forecasts can enable pilots to avoid ash areas. Ash issues were caused in 2009 by Mt Redoubt in Alaska as well as eruptions in the Northern Mariana islands, a United States western Pacific possession.

Hawaiian volcanoes are located over a Pacific 'hot spot', and they behave differently from their 'Ring of Fire' cousins.

Violent eruptions are extremely rare, and usually lava flows are comparatively gentle. In the past, glowing hot lava has devoured several villages on the Big Island of Hawaii. Kilauea volcano, in eruption since 1983, sent streams of lava into the Pacific Ocean near the Hawaii Volcanoes National Park as recent as 2009. Adjacent much taller Mauna Loa volcano erupts less frequently, but it has the potential to send lava towards some heavily-populated areas.

Although not currently destroying homes, Kilauea has been a bad neighbor for surrounding communities. Kilauea's Puu O'o and Halemaumau craters have released 1500 - 3000 tons of sulfur dioxide gas per day since mid 2008. Depending on local wind conditions this natural pollution source can reach all parts of the Big Island and sometimes the rest of the state. One by-product of the sulfur dioxide release is particulate pollution. The poor resulting air quality can burn sensitive crops, aggravate breathing problems, and greatly reduce visibility.

Tsunamis

Tsunamis are a hazard most commonly associated with the Pacific Ocean. Historical research shows that approximately 59% of destructive tsunamis have occurred in Pacific waters. While that is the case, other oceans are at risk. This fact was demonstrated by the catastrophic Indian Ocean tsunami just after Christmas in 2004.

There is a long list of very destructive tsunami events recorded in history. While the Pacific leads the list, people living in the Atlantic and especially the Caribbean region should not be too complacent. During 1867 a magnitude 7.5 earthquake occurred in the Anegada trough near what are now the United States Virgin Islands. Local tsunami waves following the quake were estimated to reach up to 40 feet high. These tossed ships on land and destroyed port facilities. That is those still left. Three weeks earlier a Category 3 (111 - 130 MPH) hurricane had damaged many existing structures.

In 1918 a similar earthquake took place off northwest Puerto Rico. The seismic waves reached at least 20 ft above sea level. I lived there 80 years later, and I could still observe remnants of tsunami destroyed lighthouses, and exceptionally sharply angled sea cliffs cut by the tsunami.

In November 1929 an estimated magnitude 7.2 earthquake shook the Grand Banks area near Newfoundland. Rushing tsunami waves entered fiord like inlets and rose more than 50 feet in a few locations. Fortunately the population was sparse in the worst affected areas. Minor tsunami effects were recorded as far away as South Carolina.

It took a long time to learn how to predict an ocean-wide tsunami. A pivotal event happened on April 1, 1946. Around 1.00 – 2.00am a very strong earthquake displaced the sea floor just beyond Unimak Island in the Aleutians. Not long afterwards a wall of water at least 100 feet high washed away the Scotch Cap lighthouse. The ocean reached an estimated 115 feet above sea level. Seismographs on Oahu recorded the event, but nobody knew that a Pacific-wide tsunami had been generated. Tsunami waves smashed into the Hawaiian Islands just as people were getting ready for work or school. The wave run up heights varied greatly around the state, but most north shores had heavy damage. The tsunami run up ranged from 30-55 feet above sea level in the most exposed locations. The water front sections of Hilo on the Big Island were demolished. 159 people lost their lives.

I lived briefly on the north shore of Oahu at Sunset Beach during 1971 - 1972. I got to know a 90-year old gentleman who often walked down a nearby path to watch the waves and surfers. One day he told me his 1946 tsunami tale. He was casting nets to catch fish when suddenly the water began to drain off the reefs. Trouble coming! He ran as fast as his then 65-year old legs could carry him towards the Koolau slopes that start 1/4 mile inland of the beach. But he could see he wasn't going to beat the incoming wave. So he climbed a tall coconut palm, and waited out several successive tsunami surges. When the event ended, there was a sea water lake left in the lowest spots near his palm tree. Eventually somebody in a boat crossed the lake and rescued the elderly yet very fit tsunami survivor.

The 1946 tsunami led to efforts to predict future events. Today the Pacific Tsunami Warning Center (PTWC) provides bulletins for Hawaii and international countries, including the Indian Ocean. The West Coast and Alaska Tsunami Warning Center covers most of the North American cost including Puerto Rico and the US Virgin Islands.

As the previous tales indicate, the usual culprit in tsunami formation is a vertical displacement of the sea floor caused by a very powerful earthquake often near the edges of tectonic plates. The 'Ring of Fire' is located on the margins of the Pacific Plate. Other tsunami causes are underwater landslides, subsurface volcanic eruptions, and if you are truly unlucky, an extremely rare large meteorite impact.

Once a tsunami is generated, it can cause major problems near the closest land, as in the 1946 earthquake. The remaining energy is transmitted as low amplitude (height) but very powerful waves, which travel at speeds of near 500 MPH in deep water. The interval or period between waves can range from a few minutes to possibly an hour. During a tsunami event several waves of varying heights are often produced, and dangerous conditions near an affected coast may persist several hours.

When the tsunami waves approach the coast they slow down and greatly increase in height. The water may be moving 20 - 40 MPH when it crosses the shoreline. Sometimes a tsunami is not very steep and appears like a very rapidly changing tide that far exceeds normal ranges. In other cases it appears as a foaming wall of water or breaking wave. The first indication of trouble could be an elevated horizon and a very deep roar much louder than conventional surf. Or if a trough arrives first, the ocean could drain well below normal sea level, exposing reefs and flapping fish. In either case you have but brief minutes to escape to high ground.

Modern tsunami research permits much more accurate forecasts, greatly reducing the false alarm rate. When seismographs detect a powerful earthquake with tsunami potential,

People running from the tsunami in Hilo, Hawaii, April 1, 1946. (United States Geological Survey)

the first job is to determine the depth. Earthquakes over 30 to 40 miles deep rarely displace the ocean surface significantly. Did the earthquake cause a vertical or horizontal displacement near the sea floor? Horizontal fault motion may generate significant earthquake damage but no tsunami.

To determine the actual wave height, tide gauges can measure long period changes in sea level that are equal to expected tsunamis wave periods. A new generation of DART buoys measure pressure changes within the ocean associated with both normal tides and tsunamis. Fortunately many tsunamis are only minor non destructive sea level changes, but tsunami height prediction techniques have plenty room for improvement.

My research indicates that the last major Pacific-wide tsunami was in 1964. By all historical measures we are way overdue for a destructive event. I hope we are ready when the time comes.

You may wonder what was the largest tsunami ever measured. Look no farther than Alaska to a partially enclosed deep fiord like bay backed by glaciers and steep slopes. One July 1958 evening a near magnitude 8.0 earthquake moved the nearby Fairweather Fault 21 feet horizontally and 3 feet vertically. A tremendous rockslide made a huge splash inside Lituya Bay clearing trees more than 1700 feet above sea level. Incredibly, two out of three fishing boats anchored a short distance away rode out waves perhaps briefly 50 - 100 feet high. The third vessel was smashed to pieces with no survivors.

METEOROLOGY & STORM SPOTTERS

This concludes the meteorology chapter for storm spotters. I hope you have learned more concerning how dangerous and interesting weather and some geological hazards work. The information you learned will enhance the quality of your reports.

Many dedicated spotters become very good amateur meteorologists in their local area. Careful observations of wind speed and direction, temperature, humidity, pressure changes, and cloud formations all provide short term weather outlooks. Doppler radar and satellite imagery can greatly expand a spotter's view.

When online some spotters enjoy evaluating numerical weather forecasting models. The model physics is far beyond the scope of this chapter. But some model displays show surface and upper level weather patterns for more than a week ahead of time.

Every model has strengths and weaknesses. The numerical equations that models use must make many assumptions or approximations of the atmosphere. These inevitably lead to errors in forecast accuracy. In general, model (and weather forecast) accuracy decreases the farther out in time you project. An average of many model runs under different assumptions is called an *ensemble forecast*. *Often this consensus is the most accurate averaged over time, but there will always be exceptions. Sometimes the weather is highly predictable, and most models give fairly consistent results on consecutive future runs. At other times, there are large differences of opinion that change or 'flip flop' over time. This means changing weather patterns and low confidence in the model forecasts.*

I believe weather forecasting is gradually improving for many reasons. But surprises are inevitable. There will always be a need for talented SKYWARN spotters and human forecasters.

REFERENCES

[1] "Rare Tornado Touches Down in Alaska", *USA Today*, August 2, 2005.
[2] A Watson, T Turnage, K Gould, J Stroupe, T Lericos, H Fuelberg, C Paxton, J Burks, "Utilizing the IFPS / GFE to Incorporate Mesoscale Climatologies Into the Forecast Routine at the Tallahassee NWS WFO", 2009.
[3] Information provided by Sky-Fire Productions: **www.sky-fire.tv**

Chapter 6

Hurricanes

In 1772 two hurricanes struck the Caribbean island of St Croix. A young man working on the island as a clerk wrote to his father in the American colonies and described the hurricane: *"Good God! What horror and destruction! It is impossible for me to describe it or for you to form any idea of it... A great part of the buildings throughout the island are leveled to the ground; almost all the rest very much shattered, several persons killed and numbers utterly ruined - whole families roaming about the streets, unknowing where to find a place of shelter... In a word, misery, in its most hideous shapes, spread over the whole face of the country."* [1]

The young man had no other family on the island and little reason to stay there except that he had no means to leave. Local residents took up a collection so that he could get an education and make a better life. The money was raised and he went to King's College (now Columbia University). In 1789 he was appointed by George Washington to be the nation's first Secretary of the Treasury: Alexander Hamilton. Hamilton's description of the 1772 hurricane would be repeated by different people, with slightly different words, many more times over the next 200-plus years; Galveston, TX 1900, Okeechobee Hurricane 1928, New England 1938, Audrey 1957, Camille 1969, Andrew 1992, and Katrina in 2005.

Of all the severe weather events that the United States faces each year few can compare to hurricanes and tropical storms. Unlike other weather events that may impact a fairly localized area, hurricanes impact large areas covering not only states but entire countries. Hurricane Katrina, for example, impacted the Bahamas, Cuba, and the north Gulf of Mexico coast. It then made its way inland dumping rain as far north as Quebec. Hurricane Ike packed hurricane force winds well into the Ohio River Valley. [2]

Amateur Radio operators who respond to hurricanes do so as emergency communicators as well as storm spotters. In this section we will treat the Amateur Radio response to hurricanes as a combination of ARES response and storm spotter response. We will look at the mechanics of a hurricane, related weather events, some history behind the Amateur Radio response to hurricanes, storm spotter reporting criteria, and then look at real world experiences from three different Amateur Radio perspectives; a resident of an area impacted by hurricanes, an Amateur Radio operator sent into an area to assist with emergency communications during and after a hurricane, and the perspective of a section manager from Hurricane Katrina.

Let's start with some basic meteorological information about hurricanes and tropical storms. The following introduction to hurricanes is written by Victor Morris, AH6WX.

AN INTRODUCTION TO HURRICANES

Hurricanes and tropical storms have long been feared as one of nature's worst hazards. Columbus encountered them in his voyages near the Caribbean. In 1635 a very powerful hurricane brought great destruction to early Pilgrim settlers in Massachusetts. Hundreds of these storms have seriously impacted the United States in recent centuries.

A hurricane is an intense low pressure area that forms over tropical or sub-tropical ocean waters. It is accompanied by a counterclockwise (in the Northern Hemisphere) rotating spiral band of heavy showers and thunderstorms. The thunderstorms may exceed 50,000ft high, and are capable of producing torrential rain. The winds are relatively light at the outskirts of the storm, but increase dramatically as one approaches the central storm core or eye. The worst winds are located in the eye wall, a ring of deep convective clouds that surround a much calmer center. An eye often begins to appear on satellite imagery when sustained (one minute

Image and path of Hurricane Katrina. (Storm Pulse, used with permission)

> **The Saffir-Simpson Hurricane Scale**
>
> The Saffir-Simpson Hurricane Scale was developed in 1971 by Herbert Saffir and Bob Simpson. Saffir was a civil engineer interested in the effect of hurricanes on structures. Simpson was, at that time, director of the National Hurricane Center. The scale, similar to the Richter scale and later the Fujita scale, assigned a number, 1 - 5, to a hurricane based on wind speed and storm surge.
>
> In 2009 the National Weather Service began an experiment to change the basis of rating from wind speed and storm surge to wind speed alone. The reason for this is that a hurricane can have a Category 4 wind speed and a Category 2 storm surge (eg Hurricane Charley 2004). This discrepancy in wind speed / storm surge was also seen with Hurricane Katrina. The new, experimental Saffir-Simpson Hurricane Wind Scale was tested during the 2009 hurricane season.

average) winds reach 75 – 85 MPH.

It is useful to define a few terms frequently used in National Hurricane Center (NHC)[3] advisories. A *tropical disturbance* is a disorganized area of bad weather that lacks a well-defined low-pressure center. A *tropical depression* has definite rotation around a circular low-pressure area and maximum winds under 39 MPM. The next step up is a *tropical storm*, with wind speeds from 39 to 73 MPH. *Hurricane* force is 74 MPH or higher. In the most extreme hurricanes sustained winds can be near 175 MPH. Note that tropical depressions, tropical storms, and hurricanes are *all* known as *tropical cyclones*.

The media often focus much attention on the peak winds of a tropical cyclone. This is certainly critical information. But a second factor, often overlooked, is the storm *diameter*. There can be huge variations in the area covered by damaging winds. Charley (2004), had a 25 mile radius of hurricane force winds while approaching Punta Gorda, Florida. Katrina (2005) and Ike (2008) brought hurricane force winds up to 125 miles from the center in their northeast quadrants.

Many factors need to come together to cause a hurricane to form. There must be an area of disturbed weather located over waters, generally 80° Fahrenheit or warmer. The atmosphere needs to be moist and unstable, allowing deep convective clouds to form and persist. There should be very little wind shear (changes in wind speed and direction) near the disturbance from the surface to 10 miles up. Wind shear tears the tops off thunderstorms, inhibiting the development of many candidate bad weather areas.

The official hurricane season is June 1 to November 30 in the Atlantic Basin, and May 15 to November 30 in the Eastern North Pacific.[4] There are rare occasions when tropical cyclones form out of season. In the Atlantic Basin the long term average season brings 10 named storms. Names are given when a tropical cyclone attains tropical storm intensity. Typically six of the storms will become hurricanes, and two or three will reach major hurricane status (peak sustained winds 111 MPH or higher). But in the past century seasonal storm totals have ranged from one (1914), up to 28 (2005).

In the tropical Eastern North Pacific area (Central America to 140° west longitude), an average season brings 16 named storms and nine hurricanes. The Central North Pacific region (140° west to the International Date Line) averages three named storms per season and one or two hurricanes. These ocean regions also have large fluctuations in storm totals from year to year. When the tropical eastern Pacific becomes warmer than usual (the El Nino phenomenon), storm totals are often above normal. Atlantic tropical cyclone tracks and frequencies vary considerably as the hurricane season passes. June storms tend to form over the western Caribbean and southern Gulf of Mexico. Although a named storm is likely only once every other year on average, some of the early storms reach land. July storms follow much of the June pattern, but the possibility of tropical cyclone formation east of the Caribbean rises as the month passes. During August hurricane formation increases rapidly, reaching a peak near September 10. Almost any warm water region from the coast of Africa to the Americas can have tropical cyclone formation. These two months bring the highest risk of United States East and Gulf coast direct hits. By October wind shear increases in the eastern tropical Atlantic and few storms form there. The majority of October systems form along a broad band extending from the NW Caribbean Sea to the SW North Atlantic. November infrequently has significant hurricane development. Most very late season systems form near the western Caribbean.

But almost every good rule has an exception. In 1985 Hurricane Kate struck the Florida Panhandle during Thanksgiving week.

Every coastal state from Maine to Texas has experienced strikes from full hurricanes. The maximum risk areas based on history are South Florida, the Outer Banks of North Carolina, and the parishes of SE Louisiana. Elevated risk regions include most of the remaining Gulf Coast and the Carolinas. Medium risk areas are NE Florida, Georgia, Virginia, Long Island, and near Cape Cod. All other Atlantic coastal locations have a slightly lower risk, but any one year may still bring a devastating hurricane.

Out in the eastern and central North Pacific, the majority of tropical cyclones move west to NW throughout the season which peaks in August. Some storms may turn NE late in the hurricane season. Cool ocean currents limit the area of 80° or warmer water. That means not too many systems survive to get far beyond the tropics. Western Mexico is an occasional target of mostly mid to late season hurricanes, and a few have struck the Hawaiian Islands. The worst storm, Hurricane Iniki brought sustained winds near 140 MPH to Kauai in 1992. Only one minimal hurricane in 1858 held together long enough to reach San Diego, California.

Tropical cyclones move along with the atmospheric winds surrounding them; most of the time these currents carry them westward in lower latitudes, although there are exceptions. As a hurricane gains latitude, it will generally encounter the middle latitude westerlies. This will cause many tracks to bend to the north then northeast, a process known as *recurvature*. The process of identifying the location of recurvature is a very challenging forecasting issue. Storms that run out of well defined steering currents may stall, loop, or do other erratic things. This makes it very difficult for hurricane forecasters to issue accurate warnings, which becomes critical if the system is near land. Once recurvature is completed a tropical cyclone often accelerates to forward speeds of 30 - 50 MPH. The rapidly approaching storm may permit only short periods of accurate warnings for anyone in harm's way.

The hurricane life cycle can vary as much as that of a human being. Some form quickly and rapidly fall apart. Others may go through several periods of weakening and strengthening. A typical tropical cyclone lasts a few days to slightly over a week. The records range from a few hours to an extreme longevity case of 31 days. This record was held by Pacific Hurricane John in 1994. He initially formed off Mexico, passed south of Hawaii and continued into the Western North Pacific, reclassified as a typhoon. Later he curved, crossed the International Date Line, and became a hurricane again. John finally lost tropical features as he neared Alaska.

Most hurricanes finally die off for one of three reasons. If they move inland, the storm loses its vital oceanic heat and moisture source and gradual weakening occurs. The second is motion over higher latitude waters too cold to sustain concentrated tropical convection. But mariners need to recognize that some tropical cyclones can transition into dangerous extra-tropical storms. The third reason a tropical cyclone can weaken is increasing wind shear often combined with the intrusion of dry, stable air into the storm circulation.

Now is a good time to examine damage and safety issues caused by tropical cyclones. Most people think of the wind first when they hear the word hurricane. The wind force exerted perpendicular to an exposed surface is proportional to the wind velocity squared. There is some debate among structural engineers regarding the wind stress calculations. A simple estimate used in the past is:

F = 0.004 x W squared
(where F is Force in pounds per square foot and
W is wind speed in MPH).

Using this approximation, 50 MPH winds create 10 pounds per square foot of force. Most people find walking difficult at this wind speed. 100 MPH wind brings 40 pounds per square foot, and 150 MPH creates 90 pounds per square foot of force.

Not only is the maximum sustained wind force an issue. Peak hurricane gusts may exceed the one-minute sustained wind by 20 - 50%. The gusts are enhanced by turbulence generated as the winds cross rough and irregular natural terrain, buildings, and other man made structures. These features are also likely to create sudden changes in wind direction. Very rapid changes in force will attack exposed structures, potentially for many hours. I hope your home is prepared for this!

Amateur Radio operators have a special concern during hurricane force winds. Every amateur with a tower and / or Yagi or similar antenna should know the engineered wind rating for his set-up. There is a risk stronger hurricane forces will exceed design parameters. In such cases one should plan to safely lower towers and take down larger square footage antennas. Complete all work at least 24 hours ahead of the first predicted strong winds.

Even vertical and wire antennas may be at risk depending on wind exposure. Any antenna can be instantly destroyed by flying debris which is not counted in wind force calculations. I suggest having a sufficient supply of simple dipole wire antennas for post-storm HF operations, plus appropriate verticals for VHF / UHF.

Tornadoes are often a surprise consequence of a tropical cyclone passage. More than 100 separate tornadoes have accompanied a few hurricane landfalls. They typically form under spiral band lines of thunderstorms. Frequently the tornadoes will develop in the right semicircle of the storm 50 - 300 miles from the center, but other locations are very possible. The twisters in tropical systems often form and dissipate rapidly. This allows local National Weather Service (NWS) offices only a very short time to issue to issue tornado warnings. SKYWARN spotter reports are critically important for these events.

Water is a major cause of tropical cyclone damage. A cubic foot weighs almost 64 pounds, and that can generate unbelievable forces when it starts moving. There are several water related tropical cyclone dangers.

Flat terrain locations can accumulate significantly large and deep areas of standing fresh water. Many tropical cyclones including depressions will drop several inches of rain per 24 hours. Wetter systems have dropped a foot to over three feet of rain in a day. This can happen hundreds of miles from the coast. In hilly locations the runoff is much faster and sudden flash flooding becomes a major concern. Flash floods can be extremely dangerous when high intensity precipitation cells repeatedly 'train' over the same location. When you are driving, remember the mantra "turn around, don't drown." Flowing water only two feet deep can sweep many autos away.

If heavy tropical rains move over a saturated river drainage basin, general major river flooding can be expected. This type of flooding is more predictable, hopefully allowing enough time for evacuations and mitigation measures such as sand bags.

Coastal low lying beach communities face the life threatening dangers of storm surge and wave battering. A storm surge is a rise of sea level higher than the astronomical tide. It is created primarily by the frictional stresses of violent onshore wind, primarily in the right semicircle of the hurricane at landfall. Many factors contribute to surge height including peak winds, the area covered by very high wind, and the amount of time the winds blow. The underwater terrain or *bathymetry* can play a significant role. All things being equal, locations with shallow bottoms well offshore have higher surge but smaller wind waves. Areas with a narrow continental shelf will have larger waves but a lower maximum surge.

It is worth noting how much difference the hurricane diameter can make on the observed storm surge. Previously referenced small peak 145 MPH Charley generated only a 6ft – 8ft surge in SW Florida. Hurricane Katrina, 120 MPH at final landfall in Mississippi, generated an all-time United States surge record. NHC investigators confirmed a 28ft surge at Pass Christian, Mississippi.

There has been a slow but consistent improvement in NHC track forecast accuracy in recent decades. This has happened due to the mix of vastly increasing computational power, improving models of hurricane physics, and better multi platform observational networks. The latter includes increasing varieties of weather satellite data, better instrumentation on hurricane hunter aircraft, buoy and volunteer ship observations, and many more real time reports from observers like you the SKYWARN spotter. At the same time in our high-tech age, a dedicated and experienced hurricane forecaster can still improve significantly on forecasts generated from computer model output.

Meanwhile there has been somewhat less success in reducing intensity change forecasting error. Hurricanes that intensify rapidly are especially difficult to identify precisely ahead of time. We know the general factors that promote rapid intensification, but there is still considerable difficulty in applying the factors to specific hurricanes. This situation can be very dangerous should a hurricane undergo rapid unexpected strengthening in the last few hours prior to landfall.

Finally, the NHC forecasters are well aware of Murphy's Law when they write their forecasts. There are unknowns that will create forecast errors. In order to play it safe, tropical storm and hurricane warnings are issued for a considerably wider area than the regions which actually experience damaging winds. The goal of the forecasters is that 95% of the time (two standard deviations for the statistically inclined) the dangerous wind speeds will occur somewhere within the warned areas. But that safety factor means that in quite a few cases you will experience conditions less severe than predicted. As technology continues to improve, it will enable forecasters to specify tropical cyclone danger areas more precisely in the future.

REPORTABLE CRITERIA

During a hurricane or tropical storm SKYWARN activation, as well as activation of the NHC station WX4NHC and the Hurricane Watch Net, is essential in getting ground truth reports to the NWS / NHC. Let's look at a couple different ways these reports are received by the NWS.

First are the SKYWARN storm spotters. Spotters in the area where the hurricane makes landfall can relay important information about flooding due to intense rain or storm surge. Rainfall amount can also be relayed. Also critical is the time the rain began and ended. This can give a good picture on hourly rainfall totals for a particular area. Damaging wind, another characteristic of hurricanes, is also important information to relay. Wind speeds, both sustained and gusts, are important information. It is possible to estimate wind speeds and these estimates can also be reported. Also report associated damage with the winds, eg "roof from house blown off" or "pine trees, approximately 75 feet tall uprooted". Damage from wind can help meteorologists determine wind speed. Remember that hurricane force winds are in excess of 74 MPH. Winds at this speed will turn many things into flying missiles! Stay inside in a safe location. *DO NOT* go outside to try and observe weather conditions when it is this dangerous.

It is also important to relay tornado activity to the NWS. As hurricanes make landfall tornadoes can be generated in the rain bands, particularly in the upper right quadrant of the storm. Getting accurate information about these tornadoes is important in issuing warnings for areas that may be impacted by them. But again, it absolutely critical to stay safe! Do not go out storm spotting if it means compromising your safety. And remember, tornadoes caused by hurricanes can appear with little or no warning. While they are typically weaker and of shorter duration they are no less deadly.

The National Hurricane Center in Miami, Florida is home to WX4NHC which works in conjunction with the Hurricane Watch Net. This net provides a vital service in relaying hurricane information to the NHC and NWS. Report information that is needed is similar to reportable criteria for SKYWARN, but with a few minor differences. The reporting station's location, along with wind speeds, both sustained and gust, are helpful. Again these may be measured or estimated. Wind direction is also needed. Barometric pressure, in inches

ARRL Alabama Section Manager Greg Sarratt, W4OZK (right), speaks with two incoming Connecticut volunteers at the Montgomery, Alabama, Red Cross volunteer marshaling center, Dave Hyatt, K1DAV (middle), of Torrington, and Dave Wilcox, K1DJW, of New Hartford. Sarratt spent more than a month processing Amateur Radio volunteers for deployment to the Gulf Coast. (Dennis Motschenbacher, K7BV, photo)

Tropical Cyclone-Induced Tornadoes

Tropical cyclones such as hurricanes and typhoons are responsible for a wide range of related weather phenomena; flooding, wind damage, storm surge, and tornadoes. Tornadoes pose a significant risk to life and property whether they are formed by severe thunderstorms or tropical cyclones. Tropical cyclone-induced tornadoes differ from severe thunderstorm-induced tornadoes in several ways. Typically they are shorter lived and somewhat weaker than those that occur in the Great Plains. Tropical cyclones also rely on warm tropical air as their energy source. As such, tropical cyclone-induced tornadoes are not usually accompanied by hail or a lot of lightning.

Tropical cyclone-induced tornadoes are most frequently produced in the right-front quadrant of the tropical cyclone in the spiral rain bands within 50 - 200 miles of the center of the storm. These rain bands can develop into severe thunderstorms that can produce tornadoes.

If we look at these tornadoes from the storm spotter's perspective, we can see how they can pose serious threats. The storm spotter's role is to provide ground truth reports to aid in the issuance of warnings. During a tropical cyclone it may be difficult or even impossible for a storm spotter to observe anything more than what they can see from their home or other secure location. The ability to see a tornado is often hindered by the tornado being rain wrapped. Since these tornadoes are often of a shorter duration than conventional tornadoes a trained spotter may not be there to observe the event. Amateur Radio operators are likely to be involved in an Emergency Operations Center (EOC) or other emergency communications operations. So where does the NWS get information on tornadoes during a tropical cyclone? Doppler radar is the primary source of information. Notification may also come in from a variety of sources but trained spotters are usually limited. Public safety officials, the news media and the general public also provide reports. Overall, tropical cyclone-induced tornado ground truth is harder to come by. This combined with the aforementioned challenges of tropical cyclone-induced tornadoes typically results in a higher warning false alarm rate.

Hurricane Ivan, which struck the Florida panhandle and Alabama in 2004, provides us with an extreme example of a tropical cyclone-induced tornado outbreak. Ivan resulted in eight deaths and 17 injuries from tornadoes alone. A total of 117 tornadoes occurred over a three-day period. The tornado damage ranged from F0 to F2. Most of the tornadoes took short tracks ranging from a few miles to about 20 miles. The Florida Panhandle experienced some of the most significant impacts from tornadoes associated with Ivan. Between September 15 and 16 the NWS Tallahassee office received 24 tornado reports and issued 130 tornado warnings, roughly a 10% higher warning false alarm rate as compared to the 2004 national average.

For Amateur Radio storm spotters tropical cyclone-induced tornadoes must be taken seriously. Despite being weaker and shorter-lived than conventional tornadoes, they still pose a significant threat to life and property. Accurate ground truth reports are critical to the warning system for this hazard. But ground truth reports are never more important than spotter safety. When spotting from an area impacted by a tropical cyclone you should make sure you follow all safety precautions. Do not go out spotting mobile when road or weather conditions make it unsafe to do so. If you are operating from an EOC or other secure location you may end up receiving weather reports to relay to local emergency management, public safety, and the NWS. Use your knowledge as a trained storm spotter to relay valuable weather information.

Damage from a tornado caused by Hurricane Ivan. (Dave Barton, AI4GF, photo)

or millibars, is also reportable along with sea state, precipitation reports, and, where appropriate, damage reports.

Whether you are relaying reports via a SKYWARN net or through the Hurricane Watch Net, don't forget the critical information: who, when, where, and how to contact. Include your call sign or name in your report. Include the means to contact you: e-mail, cell phone, IM, etc. Include your location as a physical address, latitude / longitude, or proximity to a known location, eg "I'm eight miles north of Biloxi, Mississippi on highway 15". When giving the date and time make certain you are giving the date and time that the weather event was observed. There are times where the relay of a report may be delayed.

Finally don't forget post-storm reporting. Information on storm damage is important to the NWS. If you have pictures of storm damage send them to your local NWS WFO. Many WFOs have information on submitting pictures or video on their Web sites.

THE HURRICANE WATCH NET

AND WX4NHC

In 1965 Hurricane Betsy struck the Gulf Coast of the United States. She first crossed the Florida Keys as a Category 3 before moving into the Gulf of Mexico where she intensified to a Category 4. Betsy then hit the Louisiana Gulf Coast as a Category 3 on September 9. Betsy was the first hurricane to hit the US that caused in excess of $1 billion in damages ($10 billion in 2005 dollars), earning her the name 'Billion Dollar Betsy'.

In the aftermath of Betsy several things came about. The name Betsy was officially retired from use for hurricane naming. The US Army Corps of Engineers began the Hurricane Protection Program, building levees around New Orleans (the same levees that failed during Katrina).

In the midst of Betsy's fury, the Hurricane Watch Net (HWN) was born, founded by Jerry Murphy, K8YUW. Hurricane nets were not a new thing. As early as the 1930s Amateur Radio operators in hurricane-prone areas operated nets when hurricanes threatened. But since its beginning, the HWN as been a key part of the Amateur Radio response when hurricanes strike.

Of primary concern for the HWN are hurricanes that form in the Atlantic and threaten populated areas in the Caribbean, Central America, Mexico, and the US. On rare occasions the net has been activated for Pacific hurricanes. The general criterion for net activation is a named storm that is forecast to make landfall and is with 300 miles of doing so. Hurricanes or tropical storms that do not pose a threat to a populated area and stay out to sea would not require a net activation. Since its beginning there have been 444 named storms in the Atlantic (as of 2009) and of those an estimated 75% have threatened or actually made landfall.

Currently the net consists of about 40 members that serve as net control stations. They are strategically placed throughout the US, Canada, the Caribbean, Central America, and Mexico. Net members all serve as net control. Station that report weather conditions to the net are not formal members of the HWN *per se*, but serve a critical role. Formal membership in the net is limited and highly selective. Participation in the net, however, is open and welcome to those reporting critical weather conditions related to the hurricane, particularly stations within 100 miles of the eye of the hurricane.

There are two primary purposes of the HWN; first, to disseminate the latest National Weather Service advisories on active named tropical storms and hurricanes in both the Atlantic and Pacific side of the Americas. This includes transmissions to any maritime Amateur Radio operators that may be in the affected area and, second, to gather real-time ground level weather conditions from amateurs in the affected areas and to get these reports to the National Hurricane Center via WX4NHC in a timely and accurate fashion .The HWN operates on a frequency of 14.325 MHz and utilizes SSB and APRS for relaying weather information. When necessary other bands, such as 40 and 80 meters, may be used. During net activation stations checking in are directed by net control to report certain weather conditions: station and geographic location, time in UTC, wind speed (sustained and gusts), wind direction, and barometric pressure.

We should keep in mind that the purpose of this net is for relaying information pertaining to specific weather conditions. Health and welfare traffic is handled through other nets.

So what happens to the information collected by HWN net control stations? This is where the National Hurricane Center (NHC) and WX4NHC comes in. Since its beginning the HWN has relayed information to the National Hurricane Center in Miami, Florida. In 1980 the NHC set up an Amateur Radio station, now WX4NHC, which reports could be relayed to. This was originally done by FAX and RTTY and eventually via SSB. WX4NHC, staffed by volunteers, keeps a close working relationship with the HWN. Reports that come in via the HWN are relayed to WX4NHC and then on to the NHC. The net can also serve to help keep NWS offices in touch with the NHC and, if needed, relay important traffic to NWS offices.

For more information on the NHC[3], the HWN[5], or WX4NHC[6], refer to their respective Web pages.

THE HURRICANE EXPERIENCE

Thousands of Amateur Radio operators have had direct or indirect experience with hurricanes. In recent years several hurricanes have seen enormous responses from the Amateur Radio community. The stories amateurs can tell of their experiences, both good and bad, could fill volumes. For our purposes in this book it is worth looking at the experiences of three Amateur Radio operators, each representing a different perspective. We will start with Vic Morris, AH6WX, who experienced eight different hurricanes over the course of six years while residing in Puerto Rico. We will then look at the experience of an Amateur Radio operator sent to provide emergency communications assistance during Hurricane

Gustav. And finally we will look at Hurricane Katrina from the perspective of the Mississippi Section Manager, Malcolm Keown, W5XX.

Life in the *Casa de los Hurricanes*
(The Home of the Hurricanes)
by Vic Morris, AH6WX

Puerto Rico, a United States territory for over a century, is known for lush tropical scenery, beautiful beaches, and pristine azure waters. The island also lies close to the path of Atlantic hurricanes.

I moved to Rincon, Puerto Rico on the island's northwest coast in June 1994. As a former meteorology professor I was very aware of Caribbean hurricane history. Long term records indicate that Puerto Rico is struck by sustained hurricane conditions approximately once every eight years. As an Amateur Radio operator, then KP4WN, DX was a primary interest. I wanted a near ocean location favorable for HF propagation. I picked a well built concrete house on a ridge 200 feet above sea level and 1/4 mile from the Mona Passage. I chose a three-element tribander bolted to a 20 foot mast above my roof. It could be easily lowered by two persons, an exercise in which I was to become very proficient in the following years.

Rural northwest Puerto Rico did not have the reliable infrastructure that most residents of the 50 states rely on. I lived near the end of a leaky public water supply line built in 1928. Nearly everyone in my neighborhood had water storage tanks as a back-up when the public supply failed. The electricity was also erratic at times, partly due to 160 thunderstorm days per year. Many homes and businesses installed their own power generators. The need for self reliance in Rincon proved to be essential in many hurricane seasons.

The period of 1970 - 1994 was one of below average Atlantic hurricane activity. However, there were some very destructive storms such as Andrew (1992) and Hugo (1989). My first hurricane season in Puerto Rico proved to be tranquil with just one minor tropical storm encounter. But starting in 1995 a new cycle of active Atlantic seasons began which many meteorologists feel is still in progress. Sometimes for reasons not well understood, several consecutive hurricane seasons will bring storms that follow very similar tracks. This is what happened to the northeast Caribbean island region from 1995 through the year 2000.

Everything changed in 1995. The Atlantic produced 19 named storms, the most in decades. Some of these were the dreaded Cape Verde hurricanes that are spawned off the coast of Africa. Such systems frequently become major hurricanes, taking long lived west to WNW tracks that may threaten the Caribbean or the Americas.

Early in September large diameter Hurricane Luis was just under Category 4 status (131 - 155 MPH) while nearing Antigua and other NE Caribbean islands. Hurricane warnings were posted for many islands including Puerto Rico. I decided to take my beam down even though there were signs that Luis would turn NW enough to avoid a direct hit on Puerto Rico. I experienced tropical storm force winds but no real damage. Just 10 days later Marilyn arrived on the scene from a somewhat lower latitude. I hadn't even gotten around to putting my beam back up! Marilyn, a smaller diameter hurricane, smashed into the US Virgin Islands with top sustained winds estimated at 110 MPH, just under Category 3. Rincon about 100 miles from the eye again received tropical storm force winds.

1996 was a year of unwelcome surprises. Mother Nature began to act up in early June as a minor tropical disturbance neared my QTH from the east. One late afternoon after work, I headed down to the beach to cool off. There were just a few cumulus clouds over the adjacent hills when I left my house. When I got down to the ocean a few minutes later, these clouds began to explosively build into thunderheads. Swimming during a thunderstorm is a bad idea. It is also not wise to leave sensitive electronics plugged into wall outlets most Puerto Rican rainy season (May to October) afternoons, due to the thunderstorm threat. Unfortunately I had done just that.

I quickly headed home, and started to rapidly unplug things. Suddenly a high amperage bolt hit my TH3JRS antenna. Some of the energy grounded properly, but some went through my guy wires into my roof. The remainder of the lightning flash arced over to a nearby power pole transformer. The electricity then entered my wiring, exploding one outlet and melting some wires. I was in my living room, and had just unplugged a TV when the lightning struck. It generated a blinding flash and deafening concussion much like a grenade explosion. If I had been a few seconds slower unplugging my TV, I would not be here writing this article. There was one bit of good news. My home owner's policy and ARRL equipment insurance covered nearly all the damage. It took quite a while for me to get my radio gear repaired or replaced. Meanwhile the 1996 hurricane season ramped up into more serious action.

The majority of Cape Verde origin hurricane threats to Puerto Rico occur in August and September. But the 1996 season got a head start with the Independence Day formation of Bertha SW of the Cape Verde Islands. The storm steadily developed, entering the NE Caribbean on July 7 at Category 1 force (74 - 95 MPH). Unofficial top gusts were just over 100 MPH in the US Virgin Islands. For the third consecutive threat, Rincon experienced tropical storm force winds; although there were reports of nearby waterspouts.

Hortense, the eighth named storm of 1996, approached the island of Guadeloupe with near hurricane force wind on September 7. A hurricane watch was issued for Puerto Rico. But after passing Guadeloupe, Hortense encountered an area of unfavorable upper level winds known as wind shear. Maximum sustained winds dropped to 50 MPH, and the Puerto Rican hurricane watch was lifted. Most people breathed a sigh of relief, but it was premature.

On September 9 the wind shear relaxed, and Hortense began to redevelop. The National Hurricane Center computer forecast models all indicated that Hortense would continue to track west until past the longitude of our island. But since

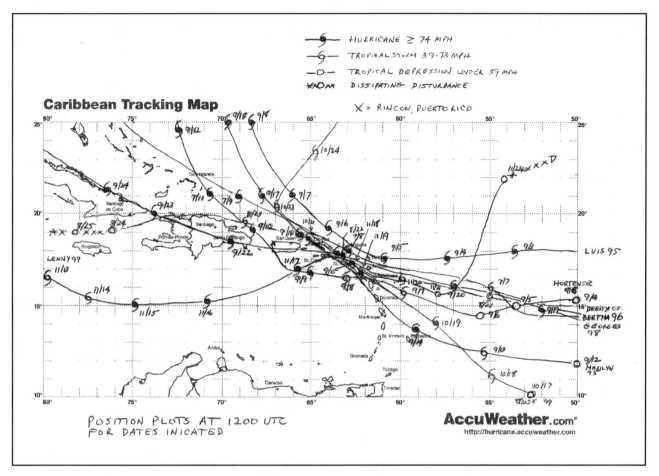

Plot showing paths of hurricanes through the Caribbean. (Source: Victor Morris, AH6WX)

Hortense was rather close to the Puerto Rico south coast, the NHC forecasters issued hurricane warnings just in case. Most people did not take the warnings seriously.

I had serious doubts about the official forecast track however, and I made full hurricane preparations. During the day on September 9 I observed thickening high clouds moving from SE to NW across Rincon. A gusty NE wind picked up but held a steady direction as the hours passed. The barometric pressure fell at an ever-increasing rate. All were weather signs known by mariners for centuries that trouble is coming.

According to the Law of Storms the wind rotates counter clockwise around Northern Hemisphere low pressure. A steadily increasing NE to east wind accompanied by lowering, thickening clouds plus a falling barometer means that one is directly in the path of an approaching storm.

As Hortense moved along, Air Force hurricane hunter aircraft fixed the storm every 12 hours. But a late day September 9 penetration was aborted due to "extreme turbulence", likely near a developing eye wall. The NHC forecasters did not know exactly where the reforming hurricane eye was, or the storm's maximum winds. Advisories issued as late as 8.00pm stated Hortense was heading west. Around 9.00pm in Rincon the steady NE wind began gusting 50 - 55 MPH, and the pressure kept plummeting. I looked to the SE and saw continuous lightning from the approaching hurricane eye wall. Now the power began to flicker. The last words I heard from NOAA Weather Radio San Juan, Puerto Rico were "Ignore the NHC advisory. The NWS radar shows that the eye of Hurricane Hortense will cross the coast of SW Puerto Rico by 10.00pm." By the time the forecast was corrected, the weather had become too dangerous for last minute outdoor preparations.

Screeching, keening winds reached hurricane force just after midnight September 10 in Rincon. My anemometer recorded four hours of frequently 75 - 80 MPH sustained NE winds. The instrument could only measure up to 105 MPH. Heavier gusts often briefly pegged the needle at speeds I estimated to be 110 - 120 MPH. After 4.00am the wind abruptly slackened and shifted to SE 25 - 50 MPH. The still incomplete hurricane eye wall was just offshore over the Mona Passage.

Even though Hortense was only a Category 1 hurricane, daylight revealed moderate to severe damage in my vicinity. Many homes owned by local neighbors were not well constructed, and a number of them had partial to complete roof failures. Numerous power lines were down, and I went two weeks without commercial power. I had to move my work computer to a house with surviving phone service in order to get Internet access. Except for that inconvenience,

I mostly made it through Hortense in good shape.

The year 1997 brought a respite from hurricanes with only one hurricane watch on a system that missed by over 300 miles. But that rest was to be a short one: then came 1998. The year began well. I saw a beautiful total eclipse of the sun on Antigua February 26. The afternoon celestial show had a backdrop of smoking Soufriere Hills volcano on the island of Montserrat 40 miles to the southwest. I felt like I was viewing the first day of creation . . . But the ancient Carib Indians as well as the Aztecs of Central America viewed eclipses as a portent of evil. I will let the reader decide.

Georges, a classic Cape Verde hurricane, formed on September 15. He quickly grew in both power and size. Sustained winds reached upwards of 150 MPH as Georges drew closer to the NE Caribbean islands on September 19. I had very bad feelings, both professional and intuitive, about Georges as early as September 17 - 18.

My hurricane research contained plots of all hurricane tracks that ever struck Puerto Rico for over 200 years during various months. Georges was moving right down the center line of all September direct hits. Early indications in the upper atmosphere showed nothing that would divert this extremely dangerous threat. Time for battle stations!

When facing a threat as serious as Georges, I prefer to begin preparations at least three to four days in advance of the potential blow. This is a good time to fuel up all vehicles and gas cans. Buy weeks worth of food that won't require refrigeration, store lots of drinking water, and charge up and / or replace weak batteries. Should the storm not live up to expectations, none of the early preparations go to waste. If you are subject to storm surge flooding, river or flash flooding, mud slides, or serious wind damage preplan your evacuation routes. Contribute to the SKYWARN and ARES networks where possible. HF real time reports to WX4NHC at the National Hurricane Center are encouraged too. These can provide invaluable data to hurricane forecasters. Weather reports should include information such as station location, time, wind direction and speed, barometric pressure, and rainfall. Include reports about flooding, storm surge height, and wave height if you are in position to provide them. But remember your personal safety comes first!

If you using a VHF repeater system, remember that repeater outages will be likely as hurricane conditions worsen. If this happens, a suggested technique is to utilize simplex operation on the repeater output frequency. This provides two advantages. One will have a better chance of finding other stations since they will be listening on this frequency. Later when the repeater comes back online, everyone will know, and they can switch over to the repeater.

Hurricane Georges maintained a relatively straight dangerous track, and I continued to elevate my preparations day by day. Fortunately a brief period of upper level wind shear weakened Georges to a still potent, large, and sprawling Category 2 hurricane as he entered the NE Caribbean. His first of seven landfalls was the island of Antigua.

In Rincon the daylight hours of September 21 provided the last chance to complete final preparations. That last day two friends arrived to help in last minute tasks. First item was to take down my beam yet again. Then we decided to guy wire a 450 gallon water tank on my roof. Following that we removed a 10 foot diameter C-band satellite dish and lashed it to 18 inch high rebar extending from my cement roof. There was only a small amount of glass window area to board up. Most of my house had *ventanas*, heavy aluminum slats that could be cranked down to keep a hurricane outside. Meanwhile the local civil defense authorities drove a bus down my street, forcibly evacuating everybody living in flimsy structures.

Just before Georges reached eastern Puerto Rico, the eye wall contracted and the clouds grew taller. The hurricane had regained Category 3 intensity. Georges took the worst possible track for Puerto Rico. In the Northern Hemisphere the strongest winds of a hurricane are generally located near the eye in the right front quadrant looking along the track that the tempest is taking. If a hurricane is moving to the west, the NW quadrant is the worst one. Georges followed an oscillating west to WNW track close to the south coast of Puerto Rico. The entire island endured a prolonged period of very destructive hur-

Destruction following Hurricane Georges in 1998. (Victor Morris, AH6WX, photo)

ricane conditions.

At sunset in Rincon my mother who lived with me asked "Where is the hurricane?" At that time the wind was NE 40 MPH, gusting to 60 MPH. That wind speed would be hardly noticed by someone who had spent half a century on Cape Cod. I replied "just wait." The power failed shortly, and I saw my first peak gust over 90 MPH by 8.30pm. From that point on there were nine consecutive hours of sustained hurricane force winds, and the gusts began to peg my anemometer ever more frequently.

The very worst conditions were during the first two hours of September 22. My anemometer cups never blew away, a testament to a well-engineered instrument. Several times as the eye edged off SW Puerto Rico I had one minute sustained wind speeds of 100 MPH. There is very irregular terrain near my former QTH, which generated very strong wind gust accelerations. When the worst gusts arrived the noise level was equal to a jet engine whine in my back yard. I couldn't hear a human voice or the debris smashing into my home. The peak rushes of wind lasted up to 30 seconds, keeping the anemometer needle pegged. I estimated some gusts reached 150 MPH. KP4MYO located 20 miles NNE of me measured a 165 MPH peak gust, making my estimate credible. During the gust impacts, the solid concrete structure shook as it might during a local magnitude 4 to 5 earthquake (Puerto Rico has considerable seismic activity too). I could also observe the needle on my barometer visibly jump around in response to the extremely sudden changes in forces attacking my house. Sometimes the wind speed varied by up to 100 MPH in a matter of seconds.

Dawn on September 22: Georges moved on to the Dominican Republic and the wind rapidly fell below hurricane force. For the first time my mother and I opened the door. She said "It looks like a bomb went off." I couldn't have described conditions any better. My street was blocked by all imaginable types of debris including corrugated tin roofs, remnants of wood homes, uprooted trees, and broken power poles. Able bodied locals acted as a *de facto* CERT team and manually cleared away enough rubble to make the street passable. I hiked around my village to document the worst destruction I had ever seen in my life. Every wooden house along my street except one lost at least a roof. Quite a few were totally demolished.

NHC / FEMA reports later indicated that 72,605 houses in Puerto Rico were damaged significantly and 28,005 were destroyed. Georges became the most costly hurricane in Puerto Rico history. But a great deal of hurricane preparations had paid off. There were no fatalities on an island with four million residents.

You may wonder how well my storm preparations worked. I had some surprises awaiting me. My 450 gallon water tank (weight 3000 pounds when full) took flight and smashed to pieces 200 feet downhill from my place. The satellite dish, lashed to the roof, was still there. But the wire mesh in the dish was blown out by the wind, making it useless. It took 26 days to restore commercial power, and a month to get municipal water. Fortunately a gas stove, a high quality generator, and extra water tanks at ground level saw me through the long hurricane aftermath.

The year 1999 brought some new surprises. A late October Hurricane Jose approached from low latitude waters east of Trinidad. Fortunately he narrowly missed Puerto Rico, and no hurricane force winds were reported. So was that it for the year? Based on climatology the odds were 99% that the season was over.

One factor in evaluating forecasting skill is the ability to anticipate and predict extremely rare events. In the second week of November a tropical disturbance slowly developed off Nicaragua. November Caribbean storms are not very common, and most that do develop head towards Cuba, bypassing islands farther east. So I initially had little concern over the system, which was to be named Lenny. In his first few days he moved in a slow loop over the NW Caribbean much as expected. Then he began a course to the ESE then east, meanwhile becoming a full hurricane. Lenny moved well to the SE of Jamaica on November 15.

NHC models consistently predicted a turn to the NE or north, and hurricane warnings were issued for Puerto Rico for the seventh time in five years. By this time the constant wear and tear of making hurricane preparations had taken a toll on me. In 1999 I correctly chose to ignore Jose. But Lenny, as a unique case, was very troubling. Neither NHC forecasters nor I had any experience with a mid-November hurricane moving east across the Caribbean. I finally chose to do nothing as the steering currents near Lenny appeared to be steady from WSW to ENE near the longitude of Puerto Rico. I predicted those would bring the eye no closer than 150 miles from my home. At the time of my decision Lenny was just a Category 1 hurricane late on November 16.

Overnight tropical storm force winds gusting to 60 MPH buffeted my place. But the wind direction gradually backed from ENE to north, indicating that Lenny was near my expected track. But after passing well south of western Puerto Rico, Lenny finally turned more NE and intensified explosively. Maximum sustained winds briefly reached 155 MPH in the SE eye wall when Lenny was tangent to St Croix in the US Virgin Islands. Never before had a November hurricane strengthened so dramatically.

After passing St Croix, Lenny slowed forward speed and lost some force. He caused widespread problems in the eastern Caribbean for several days. 'Backwards Lenny' became the first known hurricane to transit the entire Caribbean from west to east since the days of Columbus.

My tale ends on a whimper. My eighth hurricane warning came from Debby, a year 2000 storm. Although of Cape Verde origin, she only reached minimal hurricane status when close to Puerto Rico. The hurricane was asymmetric, and almost all the serious winds were confined to the northern semicircle. When the center of Debby was just 70 miles north of Rincon, it was just overcast and dead calm. I held an outdoor barbecue on my deck to celebrate our reprieve.

The eight hurricane warnings in six seasons for Puerto Rico proved to be a four decade record for any land area under NHC responsibility. A similar run of hurricanes occurred near the North Carolina Outer Banks 1953 - 1958.

I gained more hands-on hurricane experience in several years than many coastal residents do in a lifetime. I hope this course of events never repeats itself. But remember the lessons: expect the unexpected, be as self reliant as possible, and recognize the limitations of official forecasts. Since most of you are not meteorologists, I strongly urge you to err on the side of caution when considering hurricane preparations.

As a footnote, I moved to the NW side of the Big Island of Hawaii in late 2001. But Mother Nature still had a few more lessons. A 25,000 acre wild fire burned as close as four houses away from me in 2005. In 2006 a close magnitude 6.7 earthquake very nearly threw me out of bed early one Sunday morning. And my tropical 'friends' have not forgotten me either. We had hurricane watches from Jimena in 2003 and Flossie in 2007.

My First Hurricane, Gustav 2008
by Mike Corey, W5MPC

On a Sunday afternoon in August 2008 I was contacted by the Mississippi section manager with a request for communications assistance in Louisiana. Hurricane Gustav was approaching the Louisiana gulf coast and was expected to make landfall on Monday morning. An emergency operations center in St Helena Parish needed communications support, preferably before the storm hit.

After confirming with the local homeland security director and getting directions I loaded up the car and set out for the five-hour drive to Greensburg, Louisiana. Arrangements had been made with another local ARES / SKYWARN member to come down three days later to assist and relieve me. Despite being active as a storm spotter for many years and having seen tornadoes, floods, blizzards, and ice storms I had never experienced a hurricane before, so nerves were running a bit high.

Having been involved in ARES and SKYWARN for sometime I had a basic 'go-kit' ready to go. I added a few other items, an HF rig, base 2 meter antenna, and a G5RV and thought I was all set. Shortly after the storm made landfall I had a running list of 'should have brought' items.

Upon arrival I was briefed by the local homeland security director and began setting up radio equipment. It was almost midnight when I arrived and too dark to set up antennas. The forecast called for Gustav to make landfall at 9.00am Monday morning, we thought this would give us enough time to finish the antenna installation at first light. At 7.00am there was a briefing via WebEOC® with the Governor's Office of Homeland Security and Emergency Preparedness (GOHSEP). Following that last minute preparations were made. Due to a shortage of available hands a temporary HF antennas was deployed. A Ham-Stick type antenna was mounted on a cookie-sheet with some wire radials and placed on the roof of the EOC. It didn't work great but allowed us to stay in touch with GOHSEP. At the last possible minute the G5RV was put up.

Assisted by local public works employees the antenna was placed on a 35 foot tower. At the time of its installation the rain had started and the wind had kicked up to 60 - 70 MPH with gusts considerably higher. Despite being drenched we were on the air. At approximately 9.00am Gustav struck and claimed its first victim, the Ham-Stick. The winds had bent it in half and blown it off of the roof!

Not long after this the winds damaged the doors to the EOC. Until the storm let up it was a full time job keeping the water out. An hour after landfall the EOC generator failed. It wasn't until three days later that we were able to get a repairman out to fix it. To get us through until the generator could be brought back online two small portable generators were brought in. One ran the lights on the EOC floor and the other provided power to the radios. No lights and lots of water in the building presented just a few of the problems we had to contend with. It was also during the first few hours of the storm that cellular communications failed; first voice, then text.

Immediately following landfall rain was the main thing we dealt with as far as weather was concerned. In all Gustav dumped around five to seven inches in that area, while other nearby parishes received more than 15 inches. Wind was another factor, causing trees and power lines to fall blocking roads. As Gustav made its way inland reports of tornadoes started to come in, the first was received around 10.00am.

The first day we received nine reports of tornadoes in the parish. As the reports came in to the communications room they were relayed to the EOC floor and to the National Weather Service. On occasion relaying a report to the nearest NWS office took a circuitous route. Around 9.00pm on Monday evening a report came in of a tornado in the southwest corner of the parish. Attempts were made to contact NWS Slidell and Jackson, but were not successful. Contact was finally made with the Amateur Radio station at NWS Dallas-Ft Worth. Within a matter of minutes a tornado warning was issued for St Helena parish. Despite the round about means of communication we were able to stay in touch with the NWS and receive periodic four-hour forecasts.

By the end of the first day there was a lull in activity that allowed for EOC staff to rest. This little down time even allowed for a quick QSO with UA0SJ in Asiatic Russia on 20 meter CW (storm spotters are DXers too!) Not long after midnight, though, the rest came to an abrupt halt when a report came in of a gas leak. A request was made to get a message through in a hurry to the gas company to get a crew out. The situation was critical because the area with the gas leak was about to go underwater. A message was passed via the 3872 tactical net to the gas company. The crew eventually arrived and took care of the situation. Around 2.00am everything was stabilized so EOC staff could get some sleep.

The second day began with more reports of tornadoes in the area. Tornadoes caused by hurricanes are a bit different than the tornadoes usually associated with thunderstorms. Hurricane-induced tornadoes, which generally appear in the upper right quadrant of the storm, are usually short lived and weak (rarely above F1). Ground truth reports may be hard to come by for several reasons; spotters may not be able to

Storm Spotting & Amateur Radio

safely get out and observe, the tornado may be rain wrapped, and it may not last very long. Tornado warnings after a hurricane makes land fall may not be accurate because of these factors, but it is better to err on the side of caution.

The second day also marked the beginning of a recovery phase for the area. There were needs in the parish for fuel, food, water, and ice. The EOC needed the generator fixed, more 700 MHz radios, phone and Internet, and security personnel. The staff of the EOC was fairly small so many had to wear more than one hat. My role transitioned from a radio operator to assisting the local homeland security director as his deputy. Throughout day two requests were submitted for the needed commodities. Also it was a time for EOC staff to send messages out to family and friends to let them know that they were OK. The first to arrive was the Louisiana Air National Guard with a Rapid Comms trailer that provided satellite Internet and phone service. Despite numerous requests the security personnel, Army National Guardsmen, had not arrived by the end of the second day. We needed to have them arrive before the fuel, food and ice, so time was of the essence.

Day three we found the proverbial needle in a haystack, a generator repairman. After several messages were passed through the tactical net one was found about an hour away in Baton Rouge. By midday the generator was back up and running. This made an Internet connection possible again, so the Air National Guard unit with the Rapid Comms trailer could pull out to be deployed elsewhere. There was still no sign of our security team or the supplies we had requested so follow up requests had to be made. And as shelters filled up Amateur Radio was put to use relaying information for the Red Cross. Communications assistance was also given to local health department officials and social services. Day three was also the arrival of the first relief Amateur Radio operator, Doug, K5DSG.

An interesting challenge for day three came from the media. Local media were reporting that supplies were available from the EOC and people just needed to stop by. There were two problems with this. First, we hadn't received any supplies by this point. Second, if we had any supplies we would not distribute them from the EOC for security reasons. During Hurricane Katrina a similar problem occurred. A local 911 center determined that in such a situation the communications center is better suited to set the local media straight instead of going through normal public relations channels that might be tied up. We employed that lesson learned and stayed in contact with local media so they could give accurate reports. Since the EOC staff was small this freed up the director to handle other issues.

By the end of the third day our requests for supplies were being filled. Starting around 11.00pm supplies started arriving; ice, oxygen tanks, 700 MHz hand held radios, security personnel from the Tennessee National Guard, personnel to help distribute supplies from the Kentucky National Guard, bottled water, MREs (Meals Ready to Eat), and fuel.

The back side of the emergency operation center in Greensburg, Louisiana showing the tower and G5RV. (Mike Corey, W5MPC, photo)

With supplies now in hand a plan for distribution could be made.

Day four was to be our final day in St Helena, Louisiana. More Amateur Radio operators arrived to assist at the Red Cross evacuation shelter: David, KE5STF, and Peter, KE5STL. Distribution of food, ice, water and fuel also began. The plan for distribution worked extremely well. People lined up along a NW to SE road north of the distribution point. They then turned onto a road that led south to the distribution point and picked up supplies. Each person was allowed to receive a set amount of food, ice, water and fuel. They then proceeded south to another road where they could turn east and head back into town or west and head into the parish.

After passing some final messages for the Red Cross, Department of Health, and Social Services we verified that the communications situation was stable and nothing else needed. By 2.00pm we determined 'mission complete', packed up, said our farewells, and hit the road for home.

Many lessons were learned during those five days in St. Helena, Louisiana. In retrospect I am certain that we were not the first to learn these lessons. Here is what we learned from our experience.

Go in with a 'mission plan'. The chance to assist when needed is quite exciting and in many places having extra personnel is very welcome. But a volunteer that shows up to help without a plan or set of goals can quickly become a catch-all for every imaginable task. Before you go establish what your goals are, a time-frame to get them accomplished, and know your limitations. In our case we had a three-point mission plan: set up Amateur Radio communications, provide communications while assisting in re-establishing normal communications, and make sure communications are stabilized before departing. Admittedly we undertook more than what our plan outlined: media relations, EOC management, and supply requisition. Through my experience working in an EOC I was comfortable handling these tasks and at no point felt that I was outside of my ability to do the job. Not everyone would be able to do this. It is still best to stick with the plan and not vary too much from it if at all possible.

If you find yourself in a disaster such as a hurricane plan for being self sufficient for five days. Despite being in a well-equipped EOC we were without power, cell phone communication, Internet, shower facilities, and had only the supplies on hand for the first few days. Having a well supplied go-kit will help greatly in getting by those first few days. Also include in your go-kit a small portable weather measuring device that can tell temperature, barometric pressure, and wind speed. This will allow you to monitor weather conditions and report them, if needed, to the NWS.

There is a definite pattern to cell phone outages. Voice communication is the first thing to go since it requires the most bandwidth. Text messaging is about the last thing to go since it requires very little bandwidth, so if you can't make a call try sending a quick text message. The message may not go through immediately but most phones will make repeated attempts to send the message.

Training is an absolute essential for responding to

Rapid Comms trailer that provided satellite-based Internet and phone service. (Mike Corey, W5MPC, photo)

disasters. At the basic level you should have completed SKYWARN or a comparable storm spotter training course and the ARRL Emergency Communications Course. Beyond that online courses available through FEMA are of great value to disaster responders, especially EOC, working with volunteers, and NIMS courses.

In responding to St Helena we found ourselves working, in a disaster, alongside people we had never met and knew nothing about, in a town we had never been to. And from their perspective we were two strangers and they knew little to nothing about us. There is a story that is worth sharing. One of the great college rivalries is between the University of Mississippi (Ole Miss) and Louisiana State University (LSU). I was employed and attended Ole Miss and showed up in St Helena Louisiana donning my Ole Miss Baseball hat. The EOC was staffed by diehard LSU fans, and many grumbled about being sent an Ole Miss fan to assist them. Over the course of our time there a good working relationship was built and I showed that despite being an Ole Miss fan I could be trusted. Several months later while passing through the area, coming back from New Orleans, I stopped by and visited with the local homeland security director. After comparing notes on our experience during Gustav I presented him with his own Ole Miss Baseball hat. He said that is one souvenir that would get a prominent place on his desk.

During a hurricane many volunteers come from outside the affected area and find themselves in the same situation. It is absolutely critical that you have good people skills and not lose sight that you are not only a volunteer but a guest in their community. It is important that you show that you can be trusted and can work as part of the team. Keep in mind this team may have been working together for many years so this sounds easier than it really may be.

From a storm spotter's perspective a hurricane offers several serious challenges. During normal spotter activation there are mobile spotters that report conditions that they observe. During a hurricane mobile spotters most likely will not be available, making ground truth reports hard to come by. Many reports that we received came from the general public and most likely were from people not trained in observing severe weather. Contact with the nearest NWS office may also be a challenge. Lines of communication may be routed through other places, such as our case contacting the Dallas-Fort Worth NWS office. Getting an accurate forecast is also more difficult. We would get four-hour forecasts but they were far less detailed than we would have preferred. And forecasting exactly where a hurricane will make landfall is just not possible. Two hours before Gustav made landfall it could still not be pinpointed on a map.

Response to a hurricane is definitely a combination of storm spotter response and emergency communication response. If you respond and neglect the storm spotter aspect you could find yourself under prepared.

Hurricane Katrina
Ham Radio Operations in Mississippi
by Malcolm Keown, W5XX

Mississippi is in a unique climatological location resulting in a variety of weather events. In the winter frigid air from the north collides with warm Gulf flow which can result in severe ice storms. Spring brings warmer weather but the same type of weather systems with sharp temperature differentials resulting in tornadoes and widespread damaging thunderstorms. As the summer ends the Southern Coast of Mississippi is subject to hurricanes while the northern half of the state can be traversed by tropical force winds and rain. The result is that Mississippi Radio Amateurs need to be ready for emergency response the year around.

In recent times the Mississippi Coast has been impacted by two catastrophic storm events - Camille in 1969 and Katrina in 2005. Most hurricanes form off the western coast of Africa as tropical waves and later organize into low pressure systems as they move across the Atlantic. Katrina was different. This storm formed over the Bahamas on August 23, 2005 and crossed Southern Florida as a Category 1 hurricane, resulting in deaths and flooding before reconstituting over the Gulf of Mexico into a Category 5 hurricane. As the storm approached the Gulf Coast it weakened to Category 3, making landfall near the mouth of the Mississippi River, and then moving north across the marshlands east of New Orleans making its final landfall in Hancock County, Mississippi.

Due to the counterclockwise rotation of the storm, the Mississippi Coastal Counties were pounded by 120 MPH winds and a devastating 25 to 30 foot storm surge which penetrated as much as 12 miles inland. Just to the west of Hancock County, the City of New Orleans was inundated

Amateur Radio operating position showing radios for HF / VHF / UHF Amateur Radio communication, satellite telephone, and 700 MHz radio that linked emergency operations centers in the area and the Governor's Office of Homeland Security and Emergency Preparedness. (Mike Corey, W5MPC, photo)

from the north as the winds rotating around the low pressure area to the east of the city raised a storm surge across Lake Pontchartrain that pushed devastating floodwaters into New Orleans. After landfall, Katrina traversed the length of Mississippi as a Category 1 hurricane and later as a tropical storm exiting the state on its northeast corner some 400 miles to the north.

Although the ham population in Mississippi is thin, considerable preparation had been put in to motion by ARES members to deal with storm events. These hams had practiced operating under emergency conditions during Field Day, coordinated with served agencies through planning for the annual Simulated Emergency Test, regularly checked into nets to understand net protocol and how to pass formal traffic, and taken the ARRL Emergency Communications Courses to understand emergency procedures and organizational structure. In addition, Mississippi had signed assistance Memoranda of Understanding with surrounding ARRL Sections to operate a net to handle tactical traffic, leaving the traditional 'Mississippi frequency' of 3862 kHz open for the Magnolia Section Net and Mississippi Section Phone Net to handle health and welfare (H/W) traffic. This was fortuitous during Katrina in that many of the experienced Mississippi Net Control Stations (NCS) were in the southern part of the state and were off the air. Fortunately, NCS primarily from Arkansas, Louisiana, and Texas stepped up to help keep the tactical and H/W nets in operation.

By the time Katrina had passed out of the state hams were short on sleep, had been wearing the same clothes too long, had too many flashlight batteries that were shot, and had gas gauges and cans that were empty. Realizing that there was absolutely no communications on the coast, hams rose to the occasion and activated W5SGL at the Harrison County EOC and the KC5IF 146.73 repeater (on battery power) which was the only repeater still intact. Gulf Coast District Emergency Coordinator Tom Hammack, W4WLF, noted that for the first three days after Katrina made landfall, Amateur Radio was the only functional communications on the Gulf Coast.

As a result, W5SGL through KC5IF/R coordinated emergency communications across the Gulf Coast for local fire, rescue, police, and ambulance services, and arranged logistics for medical evacuations, feeding, heavy equipment movement, and military operations. In addition, operators at W5SGL expedited press releases and health notices, as well as handling traffic for the Coast Guard, FEMA, the Mississippi Emergency Management Agency (MEMA), and missing persons and disaster mortuary teams. The W5SGL log shows that over 100 trapped victim rescues as well as numerous ambulance dispatches for medical emergencies resulting from operations at the Harrison County EOC ham station.

As the situation began to stabilize other hams managed to get on the air and help with many types of tactical traffic as well as with H/W traffic. Mississippi hams to the north of the Gulf Coast managed to link the Gulf Coast by 2 meter repeater to MEMA in Jackson, providing a vital link for passing traffic. In addition to providing emergency communications, hams also assisted in the disaster relief effort in other ways including helping in shelters, participating in search and rescue operations, clearing roads and utility rights of way, and securing and delivering needed supplies. Hams were truly willing to go the extra mile in spite of family needs, enormous personal losses, and professional and job commitments.

Humanitarian assistance reports due to the efforts of hams were numerous. One particularly interesting report came from Forrest County Emergency Coordinator Lex Mason, KD5XG, of Hattiesburg, who worked at the Forrest County EOC during Katrina. The request that really stuck in his mind was a call from Camp Shelby, a National Guard Training Base, just south of Hattiesburg. A soldier in Iraq called Camp Shelby officials and requested help in contacting his father, who lived by himself and, if not found, would die for lack of medical care. Lex was very concerned about the soldier's dad and the remote area where he lived. The directions included going to Lucedale, Mississippi, turning south on Highway 57, crossing the Jackson County Line, and then turning west. Additional instructions were given about passing a cemetery and church. He lived on a dead-end road somewhere in the woods past the church, all very confusing to say the least. Through the directions received, Lex surmised that the man was probably in Jackson or Harrison County. The devastation in these two counties was unbelievable. There were no communications in the area and many roads were impassable due to downed tress. Not knowing which county the man lived in added to the difficulty.

Fortunately, the soldier requesting the assistance later gave a Perkinston, Mississippi address for his dad, which would put him possibly in Stone County. KD5XG was able to make contact with Sarah Purvis, KC5NFI, in Stone County. She delivered the message to the Sheriff's Office. Within two hours Sarah reported that the dad had been located and was safe. The contact at Camp Shelby was advised that the soldier's father had been contacted and was OK. Lex said that helping this one soldier locate his dad and giving him assurance that he would receive the necessary medical care was

Total devastation in Waveland, Mississippi. (National Weather Service)

worth all the work devoted to the Katrina Emergency Operation. He further said that our Military Service personnel put their lives on the line for us daily and helping one of them was indeed a privilege.

After Katrina had passed and Mississippi hams had a chance to get their lives back into order, a number of them met in Hattiesburg to discuss the lessons learned from this catastrophic event. These lessons were summarized as follows:

(a) Government and non-government agencies should not attempt to activate any ham radio emergency communications response without proper ARRL Section-level ARES coordination. This is not to say that emergency communications assistance was not needed during Katrina; however, there was no ham radio national / regional emergency response plan in place, thus the uncoordinated inundation of the Gulf Coast and South Mississippi with assisting operators was unexpected and unprecedented. In that current ARES policy is to organize emergency communications response on a county / district / section basis, the uncoordinated action resulted in cases of confusion, inefficiency, and hard feelings. Some outside operators seemed to think they were in charge of local ham radio personnel and resources (repeaters) regardless of the fact that an ARES emergency response structure was already in operation.

(b) A pre-populated database of emergency communications volunteers and rapid response teams (RRTs) should be maintained. The well meaning on-line method used during Katrina resulted in enlisting many helpful operators; however, many operators showed up for assignments who were ill-trained, unprepared for a stressful assignment, and had personality problems not compatible with the disaster environment, ie we don't need 'fire engine chasers' and 'cowboys'. This database should be actively updated. This database should be populated by nationwide ARES Section-level leadership recommendations, and these recommended responders must be properly credentialed.

(c) Emergency Coordinators should be in charge of ARES operations in a given county / jurisdiction provided the EC is properly coordinated with the county emergency

Views of inundated areas in New Orleans following breaking of the levees surrounding the city as a result of Hurricane Katrina. (National Weather Service)

management agency (EMA) director. In most Mississippi counties the EMA Director works under the county Board of Supervisors. The EMA Director appoints a communications officer, which may or may not be the EC. In any case, the EMA Director should be aware of all emergency nets in operation in his jurisdiction including an ARES net. In the event the EC is also the Communications Officer, he should be aware of all net operations in a given jurisdiction and be prepared to report the current status of these operations to the EMA Director at any time. There is no problem with having separate ARES, Red Cross, Salvation Army and Baptist Men's Kitchen nets in operation: the key point is that they should all properly be coordinated with the Communications Officer so that communications resources are efficiently utilized.

During the course of the Katrina landfall and following response, an article in the *Wall Street Journal* referred to ham radio as a "hobby" similar to stamp collecting. Clearly the emergency communications support provided by Amateur Radio operators during Katrina demonstrated that we are truly a "service" and not just a "hobby."

REFERENCES

[1] Alexander Hamilton, "A Few of Hamilton's Letters" (1772), The MacMillan Company, London, 1903.
[2] R Knabb, J Rhome, D Brown, "Tropical Cyclone Report, Hurricane Katrina", National Hurricane Center, December 20, 2005.
[3] National Hurricane Center Web site: **www.nhc.noaa.gov**
[4] National Hurricane Center and Central Pacific Hurricane Center.
[5] Hurricane Watch Net Web site: **www.hwn.org**
[6] WX4NHC Amateur Radio station at the National Hurricane Center: **www.fiu.edu/orgs/w4ehw**

Chapter 7

Storm Spotter Activation

Ask any veteran storm spotter and they will tell you that there is more to storm spotting than simply responding to severe weather. The storm spotter first has to be *prepared* for severe weather and this happens long before the storm arrives or even before severe weather threatens. We have discussed some of the ways to prepare: training, equipment, safety awareness. During SKYWARN activation we often face challenges and issues that we normally would not face. And often, during activation, we are pulling double duty with an ARES or RACES group. But once the severe weather has passed our duties as storm spotters are not necessarily over. There are post-storm damage reports to be made, After Action Reports, reports to the Emergency Coordinator (EC) or perhaps District EC, Section EC (SEC), Section Manager (SM) and ARRL to be made, debriefings and possibly more work to be done assisting in the aftermath. There is a lot that goes into being a storm spotter and probably even more as an Amateur Radio storm spotter. In this chapter we will cover some of the aspects of storm spotter activity before, during and after activation. Instead of going step by step through an activation process we will cover some key elements to storm spotter activation. Generally these are the elements that can be applied to almost all Amateur Radio storm spotter groups.

We will start with activity that occurs before storm spotters are activated. We will cover getting ourselves ready for activation. From there we will walk through the days leading up to a storm: making sure equipment is in order, following forecasts and hazardous weather outlooks, communication prior to activation and understanding the roles we may play once activated.

Then we can look at issues that occur during activation. Since storm spotter activity must respond and adapt to changing weather conditions we won't get overly specific about what happens during activation. Instead we will look at some critical issues that storm spotters may likely face: clear defined roles, storm spotter and SKYWARN nets, double duty between SKYWARN and ARES / RACES, lessons we can take from emergency management and handling problems with storm spotters.

In the post-storm phase we will look at how the storm spotter can follow up on their activity. We will look at gathering storm damage reports, writing After Action Reports, submitting reports to the EC / DEC, SEC, SM and ARRL and debriefing.

THE PRE-STORM PHASE

Get Into a Daily Habit

The first thing we need to do as volunteer storm spotters is to get into a daily habit of looking at forecasts and weather conditions. Many days may start out with sunshine and clear skies or no signs of severe weather, but over the course of the day may change to stormy conditions. We should monitor forecasts on a daily basis so that we aren't caught off guard. So where do we get the information and what do we look for?

The best starting spot is by going to your local NWS Web site and accessing the seven-day forecast for your area. This will give us a sense of what to expect in the short term. If there is no indication of forecast severe weather then we may not need to check in again during the course of the day.

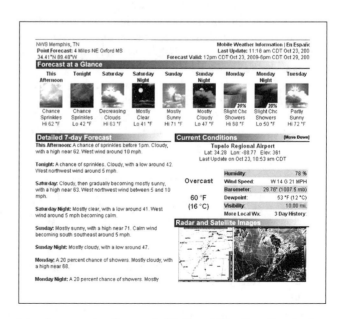

Local seven-day forecast. (National Weather Service)

If there is an indication that we should expect severe weather then we should investigate the forecast further. There are two sources of information that we can use to get more information about the forecast. First is the *hazardous weather outlook* (*HWO*). The HWO gives additional information on severe weather and the area it is expected to impact. It can cover a timeframe from the day of the weather event up to a week ahead. It also will let us know if storm spotter activation is expected or if specific reports are needed.

Two other sources we can turn to when the forecast calls for severe weather are the Storm Prediction Center (SPC) and the National Hurricane Center (NHC). Both centers are part of the NWS. The SPC has several forecast tools such as upper air maps, soundings analysis, mesoanalysis and fire weather composite maps that we can look at and get a better idea on what to expect in the short term. The NHC provides information on tropical storms that are active in the Atlantic and Eastern Pacific. While tropical storms can be difficult to predict we can look at maps showing possible tracks over a three to five day period and whether the storm is likely to intensify or weaken.

There are, of course, other sources for weather forecasts. Resources such as television and radio weather broadcasts, newspapers, weather Web sites such as AccuWeather and The Weather Channel all provide good short term forecasts. However we go about getting the short term forecast each day it is a good daily habit to develop. Taking a look at the NWS seven-day forecast and perhaps following it up with a look at the SPC or NHC information takes only a few minutes and gives the storm spotter valuable insight on what may lie ahead in their area.

There is one thing about forecasts we must keep in mind. If we look at the local NWS site or a similar forecast resource we notice that it will generally provide a forecast for seven days. Weather can be difficult to predict even day to day let alone days in advance. Forecast technology has made it possible to predict weather several days in advance - but not with 100% accuracy. Most of us can recall times when severe weather struck with no warning. Many weather sources have available 10-day forecasts. We must remember that the further out the forecast goes the less accurate it will be.

Software

In an earlier chapter we discussed software. Many storm spotters use weather software to monitor weather conditions either from home, mobile, or from their cell phones. These programs are great tools to use to see what it heading our way, current conditions, view watches and warnings and integrate data such as APRS or CWOP. However, since most days we are not going to encounter severe weather we may go weeks at a time without having to fire up the weather software. Since this software is such a vital tool to the storm spotter we cannot afford to become rusty in using it. We need to look at it the same way we look at the radios we use. When we get on the air and make contacts we are, sometimes unknowingly, developing the skills we will use during severe weather or some other emergency. When severe weather isn't expected or present we can take this time to fire up the software and learn more about its use. Go to radar sites that are indicating severe weather and try out different software features and practice radar interpretation. This is also a great time to make sure your software has been properly updated and to fix any bugs.

Vehicles

Proper vehicle maintenance is important at all times. For storm spotters there are some things that should be checked before activation. Make sure windshield wipers and tires are in good condition, top off the fuel tank, double check all emergency supplies (first aid kit, fire extinguisher, blanket, jumper cables, etc) and all communications equipment in the vehicle is working properly. Also make sure all lights are working properly: headlights, marker lights, turn signals, brake lights and interior lights. And make sure all routine maintenance is up to date.

Mobile storm spotting can be quite risky, more so in a vehicle that is not up to the task. By including a vehicle inspection in your pre-activation plan you are taking steps to stay safer while on the road in inclement weather.

The Storm Kit

One item the mobile storm spotter can have ready ahead of time is a 'storm kit'. Most of us are familiar with go-kits designed for either short term (less than three days) or long term (more than three days) use. The storm kit is similar in that it is a grab and go kit, but serves a different purpose. It is designed to have the tools a storm spotter needs ready to go and small enough to keep in the vehicle with them.

What do we put in our storm kit? Much like what to put in the go-kit the answer depends on the individual and what they may respond to as a storm spotter. So let's take a look at a sample storm kit. Since it is going in the vehicle with us we don't want it taking up too much space or make it difficult to find what we need in a hurry. Some items like batteries, repeater directories, spare parts for the handheld will be left to the go-kit inventory.

One item that we might keep in the kit is a small camera. As we discussed in the equipment chapter a small digital camera can be a handy tool for the storm spotter. And many of today's cameras also have the added feature of capturing video. Many cell phones have these features too but the trade off is that they are generally lower resolution and can miss some detail. So even if your cell phone has the ability to capture still and video images, keep a regular camera in the bag. A video camera is another option here. Many can capture still images as well as video at high resolution and aren't much larger or heavier than a still camera. The thing to keep in mind though is that these items rely on batteries. You will need to make sure that batteries are charged and ready to go when the kit is not in use. Your local NWS office is always very appreciative when they receive images or video of severe weather phenomena that they can use in future education and training.

A small GPS unit is another item that can be kept in the bag. Of course this may depend on whether or not you have GPS in your vehicle already. As a back-up to GPS you should also keep in the bag a good map of the area you spot in. Leave the road atlas for vacation and take along a topographic map of your area or a state atlas.

A handheld weather device such as an anemometer / thermometer would also be useful to keep in the bag. Having this available while storm spotting will allow you to report fairly accurate wind speeds and take out the guess work. And additionally it will help prevent reports of wind speeds that do not meet the reportable criteria. But much like the cameras this too requires batteries that require attention when the kit is not in use.

You should also have a blanket in your kit. During winter weather it can be used to keep warm if your vehicle gets stuck or dies and in warmer weather it can be used for protection from hail or debris from high wind. Remember though to keep it inside the vehicle, not in the trunk.

And a final item is the paperwork. Most of what you need can be kept in a three-ring binder in your kit. At the least it should include notepaper, SKYWARN guide, your local SKYWARN or storm spotting manual, net and repeater information, NOAA Weather Radio frequencies, important phone numbers and a map of your local NWS coverage area. They key is to make sure it is well organized and therefore that it is easy to find the information you need. Include with this pens, pencils and a marker.

There are other items that you will use while spotting that may already be in your vehicle: cell phone, charger, radios and perhaps GPS. The idea for the kit is to put in a grab and go bag those items you will need while mobile storm spotting. Imagine a carpenter arriving at a job site without a hammer. You don't want to respond to severe weather without all the needed tools.

Making sure you have what you need is not just for the mobile storm spotter. If you operate as a net control station or storm spot from a fixed location you want to keep certain items on hand. Make sure you have maps of the area. Computer programs such as *Google Earth* can provide great maps, but if the power goes out you might want to have paper maps as a back-up. Also don't forget to keep water and snacks on hand. Net control stations may spend many hours on the air before, during and after the storm. And have your SKYWARN manual, net control procedures and storm spotter guides handy.

Make SKYWARN Part of Your Plan

One of the duties of your local Emergency Coordinator is to "Develop detailed local operational plans with 'served' agency officials in your jurisdiction". When a disaster strikes we do not respond in an uncoordinated, unplanned, haphazard manner. If we did the response could end up being just as a

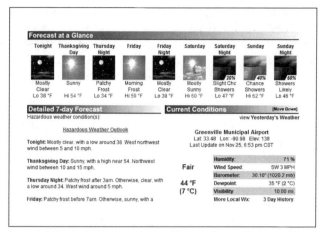

Local seven-day forecast and Hazardous Weather Outlook (National Weather Service)

big a disaster as the disaster itself. Instead, long before an emergency occurs we develop operations plans. This is a team effort between all involved, Amateur Radio, emergency management, local public safety officials: basically it is a plan on how Amateur Radio plays a part and in what way during and following a disaster. Our involvement in SKYWARN can be included in emergency planning.

The Emergency Response Plan (ERP) is a document developed by counties, cities and other local jurisdictions that describes "how citizens and property will be protected in a disaster or emergency"[1]. The production of an ERP is coordinated by local emergency management, but all involved parties have input in the process of its development. This includes Amateur Radio and SKYWARN. The ERP document contains the following: introductory information, the basic plan, annexes and appendices. The annexes and appendices are usually where our role is found. The annexes explain how the community will carry out broad functions during an emergency, such as warning or resource management. An appendix is a supplement to an annex that provides information on how to carry out a specific function in the face of a specific hazard.

One annex that is generally always found in an ERP is one that covers warnings. This annex establishes a warning system within the jurisdiction that is capable of delivering adequate and timely information to officials and the public in the event of a threatened disaster. If we think about the systems in place capable of providing warnings to officials and the general public we realize that there are more than one: the news media, the NWS, National Warning System (NAWAS) and warning devices such as tornado sirens and public alert systems. SKYWARN is a form of warning system. SKYWARN often relays reports to the NWS via a local emergency operations center or emergency manager. This provides a means to alert local officials. Some members of the public monitor our activity on a scanner and can be alerted of an emergency that way, while others will receive warnings from the NWS or media that may be the result of information that we called in.

Chances are your area has an ERP. To find out if Amateur

Radio and SKYWARN are included check with your local Emergency Coordinator or emergency management agency. If they are not part of the plan it may present an opportunity to get Amateur Radio and SKYWARN included. ERPs are not stagnant documents. There is a life cycle to ERPs. The first step is hazard analysis, then ERP development, then testing the plan and finally maintenance and revision. SKYWARN is an important part of a community's emergency plan. If your group is not part of it, it should be.

Leading up to Activation

We've looked at a few things that a storm spotter can do to prepare themselves for SKYWARN activation. Now let's look at what we need to do in the days leading up to activation. Of course there are times when SKYWARN activation is needed with little advance warning. Most times, though, we will have some advance warning through HWOs from the NWS. Let's use as an example a forecast that calls for a chance of strong thunderstorms later in the week.

The HWO is the key to staying informed on approaching severe weather. The HWO is issued by the NWS and contains information on potential severe weather threats. Typically it will contain date and time issued, affected areas, timeline of potential threats and information on storm spotter activation. The HWO is issued by the local NWS office. As weather conditions change the HWO may be updated with new information. Unlike watches, warnings and advisories, though, there is no announcement made if a HWO is updated or cancelled so you have to check back regularly to keep up with changes. So how do we get this information? How do we incorporate it into local storm spotter planning?

We first start with our daily habit of checking the forecast. If you use a source other than your local NWS office Web site you may miss an issued HWO, so again it is probably best to go to the NWS for forecast information, or if other sources say there is a chance for severe weather. Now let's go to the example of the thunderstorm forecast.

First we are going to go to our local NWS Web site and select our area by clicking on our county on the map or by entering our zip code in the search bar. Each NWS site may be a little different, but it shouldn't be too difficult to locate your local forecast information.

At the top of the screen we see the seven-day forecast and below that any watches, warning, advisories, or hazardous weather outlooks.

We can see an HWO has been issued and can access it. We can see what to expect today and in the coming days. We can also see when spotters may be needed.

Once we have this information we can pass the word to our local storm spotters. This can be done by e-mail, on our local net, text messages, or phone calls. With this advance notice spotters can start making preparations for activation. After the initial HWO we need to make sure to check back regularly for updates.

Many local NWS offices utilize conference calling or NWS Chat to keep partners such as public safety officials, the news media and others updated when severe weather is expected within the following 24 hours or so. SKYWARN storm spotters may be able to utilize this. Check with the Warning Coordination Meteorologist at your local NWS office to see if you can participate.

So let's say we've followed the HWOs that have been issued over the course of two to three days and now it is looking like there is a risk of severe thunderstorms, hail, flooding and possibly tornadoes in our area. The local NWS office informs their partners via conference calls and NWS Chat. If via conference call, at the scheduled time you dial into the conference sever. Be sure to mute your line so extraneous noise does not interrupt the call. During the conference call information is given about the approaching severe weather, risk potential for different areas within the NWS office's area and perhaps storm spotter information. A Web site may also be utilized to present graphical information. The call may conclude with a chance for participants to ask questions to the forecaster conducting the conference call.

Depending on local activation procedures this may be a good point to get in touch with other groups and agencies we may work with during activation. We may also want to determine a net control schedule or rotation, how many fixed and mobile spotters we may have and double check radios, computers, the net control station, vehicles, etc and make certain that we are ready for storm spotter activation. In the hours ahead of the approaching severe weather we will want to start keeping an eye on radar. Weather systems can change course, although generally not drastically. A storm system may appear to be heading our way. By monitoring

Hazardous Weather Outlook text (National Weather Service)

it we can easily notice trends: gradual change in course, strengthening or weakening, new cells developing and read storm reports from areas it has impacted. This information will help us make the decisions leading to local activation of storm spotters.

At this point we must look at who makes the call for activation. The local NWS office will make it known that storm spotter assistance / activation may be needed, but generally they won't make the call to activate storm spotters in a particular area. They may request a SKYWARN net be activated: this would definitely be the case when primary SKYWARN net control is located at the NWS field office. The activation of local spotters is primarily a local responsibility. This can be made by the local emergency management director, Emergency Coordinator, local SKYWARN coordinator, ARES leader, or whoever may be designated to make that call. In some cases a formal activation of spotters at the local level may not happen. This does not mean an individual spotter cannot be active and submit reports. Volunteer storm spotters, whether as a group or individually, are self-activated.

Let's take a look at two types of local spotter activation: group activation and individual activation.

STORM SPOTTER ACTIVATION

The local storm spotter group may be formed in a variety of ways. It can be a formal SKYWARN group consisting of Amateur Radio and non-Amateur Radio members. It can be organized through local emergency management. Or it can be a loose association of trained spotters. There are three key characteristics regardless of how the group is formed: they are trained storm spotters through the SKYWARN program, they are organized in some way and they have the ability to communicate with each other and with the local NWS office. For our discussion we are going to look at a local group of Amateur Radio operators serving as SKYWARN volunteer storm spotters.

The first step in activation, a request for assistance, comes from the local NWS office. Before severe weather arrives the local NWS office, if equipped, may activate their Amateur Radio station to serve as the primary net control station for SKYWARN. At the local level the decision is made prior to the arrival of expected severe weather to activate the local SKYWARN net and place it in stand-by mode. So there possibly exist two different nets during a severe weather event: the NWS SKYWARN net may be active, taking reports from stations throughout the NWS CWA (County Warning Area) and the local net taking reports from local storm spotters to be relayed to the local NWS office. At the NWS level the net control station may be manned whenever there is a severe weather watch in the CWA or when weather conditions are such that activation may be likely. At the local level placing the net and spotters in stand-by mode should be done in a similar fashion, with enough lead time to allow storm spotters to make preparations. At this point we would begin making calls by radio, telephone, e-mail, pager, etc to local storm spotters to inform them of a pending activation. We

Rob Indrebo, KB9SDF, providing communications assistance following a tornado in Ladysmith, Wisconsin. (Jim Staatz, KG9RA, photo)

would also get in touch with local served agencies such as public safety or emergency management to notify them of our status. This is also a good point to make sure that our lines of communication with the NWS office are working. Remember to have back-up lines of communication: don't rely on only one method of staying in touch with NWS. During this stand-by mode we can begin to determine how many mobile and home-based spotters we have, relay stations, net control stations, EOCs, etc that we have when we go fully active with the SKYWARN net. Don't forget during stand-by mode to continue to monitor NWS announcements, HWOs, radar and conditions in areas already impacted by the storm. Critical information should be relayed to stations that have checked into the net during the stand-by period.

A common problem SKYWARN groups encounter in the period of time leading up to activation is spreading the word to members that activation is imminent. There are several reasons for this problem: time of day (early morning, during work day), members not near a radio, not enough SKYWARN members, repeater issues, etc. Today we have more means of communication than ever before: telephone, both landline and cellular, pagers, e-mail, text messaging and, as Amateur Radio operators, radio. When we need to get the word out we need to exhaust all possible means. Local SKYWARN coordinators can help alleviate some of these problems by taking several steps long before severe weather strikes. First keep updated records of all local storm spotters. This should include name, call sign, Amateur Radio capabilities, e-mail address, home and cell numbers, instant messenger address and pager numbers. Also include best method for different times of the day. Contact by work phone or cell phone may be best during the day, radio in the evening and home phone at night. Don't rely on one method. Local SKYWARN coordinators should also make sure that there are enough training opportunities throughout the year. More training means more and better trained spotters. And coordinators should have plans on what to do if repeaters fail. Always plan to have back-ups!

Once the severe weather has entered our response area,

which can be defined as city, county, metro area, etc, we can move the net into active mode. We may continue to get more storm spotters checking in, but the net is ready for business. Now let's take a look at the local SKYWARN net and the area wide SKYWARN net that may be operating from the NWS office. And we will take a look at the roles that play a part in the SKYWARN net and process of submitting storm reports.

The Wide Area SKYWARN Net

The wide area net, which may or may not be active during a local activation, is the Amateur Radio station that operates during a weather emergency from the local NWS office. It is staffed by local Amateur Radio operators who are not only trained SKYWARN storm spotters but often given additional training in net control operations and NWS procedures. The wide area net may be conducted from the NWS office or at a separate location where information is then relayed to the NWS office. The local NWS office staff, in coordination with the assisting Amateur Radio operators, determines the conditions and time frames necessary for SKYWARN net activation at the NWS office.

Since this net may cover a very wide area, the net most likely will not be conducted on a local repeater, unless of course the repeater is linked in some way, eg by *Echolink*, *D-Star*, wide area repeater network, etc. So to accommodate local SKYWARN nets and storm spotters across the CWA, HF frequencies will probably be used.

The purpose of the wide area net is to collect reports gathered either from local SKYWARN nets or SKYWARN storm spotters and relay them to the NWS forecasters. Having HF capability during events that impact a very wide area (more than one NWS CWA) also allows information to be passed on to other NWS field offices if needed. For example, a tornado touches down at the southern edge of a NWS CWA and is reported to a local net by a storm spotter. The local net can relay this information to their wide area SKYWARN net. Since it fell on the edge of the CWA it may be an issue of concern to the NWS office to the south. This information can be relayed from one wide area net control station to the other, assuming both NWS offices have Amateur Radio stations.

The Local SKYWARN Net

Ideally the local SKYWARN net should be conducted like the wide area net. The local net can operate from just about any location, but typically it is operated from a home-based station or a station located in an EOC or similar facility. In some places net control may also be the link between storm spotters and local officials. The storm spotters relay reports via net control that then relays them to the NWS and local officials. And local officials, as well as the NWS, may use net control as a point of communications with local storm spotters. The critical part to local net control is that they need to be able to stay operational throughout the activation or have a plan in place to transfer net control duties to another station if they are forced to go off the air. Loss of power, telephone, or Internet can seriously hinder net control duties. But we also must keep in mind that the weather itself may force net control to shut down. We don't want net control to stay on the air if it is not safe.

The amateur or amateurs responsible for net control operations, like their counterparts at the NWS office, have a tremendous responsibility on their hands. They should not only be trained SKYWARN storm spotters but should also have training in emergency communications, traffic handling and net control operations. During severe weather net control can be quite busy and it can be a very stressful job with a lot to do. Net control is not only responsible for the general management of the net but also often has the task of relaying critical weather updates, knowing where their storm spotters are at and help keep them out of harm's way, receiving and relaying weather reports to the NWS office or wide area net control and serving as local Amateur Radio point of contact for NWS, emergency management, EC / DEC / SEC and anyone else that may be involved in severe weather response. The demands placed on net control stations can be great, but it can be a very rewarding experience.

Local net control can be greatly helped out by the use of relay stations or assistant net control stations. These stations can be tasked with handling the relay of information to NWS, passing weather related information to severe weather nets in neighboring counties, or be used to monitor weather updates coming from the NWS. Having this kind of assistance in place can allow the net control station to focus on taking care of the storm spotters that are checking into the net. These assistants can be stations at locations different from net control, or there can be another amateur that works alongside the net control operator to handle these tasks.

During net activation the net control station must be sure to keep track of what is going on. Net activity can include incoming and outgoing traffic from fixed and mobile stations, the NWS, the wide area net, neighboring SKYWARN nets, telephone and the Internet. If there are mobile spotters, net control will need to know where they are in relation to the storm and keep them posted with current weather information to keep them safe. Net control will want to be certain to keep careful record of all these activities. We have to go beyond pen and paper methods. A computerized log can be a tremendous asset. Along with activities, be sure to log dates, times and all stations involved in the activation.

Now that we understand the types of nets involved in a storm spotter activation we must take a look at the role the storm spotter plays.

Role of the Storm Spotter

The storm spotter, the vital link between what is happening on the ground and the NWS, can operate in two different ways, mobile or fixed. Both of these have advantages and disadvantages. And during activation it is ideal to have plenty of spotters active from different vantage points. There are some things that spotters, whether fixed or mobile, must keep in mind during SKYWARN activation.

First is your number one priority, safety. Home-based spotters should make sure that they are safe to stay on the air. They must make sure there is proper grounding of all equipment in case of lightning strike. Care must be given to make sure the home or structure that they are in and all inside are safe from severe weather. They must keep alert of approaching severe weather and changing conditions.

Before mobile spotters head out for SKYWARN activation they should first make sure their family is safe and their home is prepared for severe weather. Do not head out until this is done. No-one's safety should be compromised. Mobile storm spotters must stay extremely alert for changing conditions. Net control can help keep both spotters posted on important weather updates. If it is not safe to be on the road because of the weather don't go out. During some severe weather events a state of emergency or curfew may be declared, limiting who can be on the road. These must be followed. What a storm spotter does is a valuable service, but we are not above the law or safety rules.

Let's take a look at how both these types of storm spotter function when SKYWARN is activated. We will specifically look at the purpose of storm spotters during activation, some guidelines on how they should operate and the advantages and disadvantages of both.

Regardless of whether a storm spotter is operating from home, mobile, or any other location, their purpose is the same: to relay important weather conditions that meet set criteria to an established local reporting point (net control, emergency management, etc) or directly to the NWS. The criterion that must be met is established by the NWS. In Chapter 5 we discussed different types of weather conditions and generally what the reportable criteria for each type are. For the purpose of this portion of activation discussion we are assuming our local spotters are operating through a local SKYWARN net and since the net is active it is far easier to allow net control to pass on reports than to attempt to do so yourself from home and especially mobile.

There are a couple of key things to remember about calling in reports. Only call in reports that meet the criteria. Do not call in a report that it is raining where you are (unless it meets the rainfall criteria to be considered heavy rain) or that there is lightning. Lightning is not reportable because it is so common and the NWS has access to lightning data. The spotter should carry with them their SKYWARN manual or a card / sheet with reportable criteria listed.

We also must keep in mind the purpose of the SKYWARN net: to collect severe weather reports from storm spotters. This net is not a round-table discussion or the place to generally talk about the weather. Radio traffic should be limited to weather reports, traffic related to safety, welfare and emergencies and traffic directed solely to net control. If you have to contact another spotter or a station not checked in to the net, first get net control's OK to move off frequency and second make the contact on a frequency not being used by the net. Once you are back on frequency let the net control station know. And, along similar lines, advise net control if you have to step away from the radio, will be temporarily off the air, or have to go QRT. If for some reason you are off the air suddenly, such as a power failure, make every possible attempt to let net control know, or another station checked into the net, so they know what happened and that you are OK. This is especially true if you are operating mobile and suddenly have to go off the air.

Home and mobile spotters should follow the directions of the net control station. Many times net control is acting on instructions from the NWS, local emergency management, or possibly local public safety. Just as much as spotters are the source of weather information, net control stations are the source for a wide range of information that spotters must be aware of. Net control stations may pass along information on new approaching weather, requests from the NWS or emergency management, information on roads being closed, or if a state of emergency has been declared.

Home-based spotters often have the advantage that they can provide assistance to net control with other tasks. Home-based spotters are reporting what they observe from a fixed location. While doing this they may be able to assist net control by monitoring NOAA Weather Radio, maintaining contact with the net control station at the NWS WFO, relaying information to neighboring SKYWARN nets, or by monitoring radar. The amount of additional assistance the home-based spotter can offer will of course depend on their ability to multi-task and only the individual spotter knows their own capability to do this. But remember to not take on so much that you neglect the first rule, safety.

One of the advantages of a mobile spotter is the ability to set up and observe weather conditions from a vantage point selected by the spotter. This can make it easier to spot specific weather events, follow a storm as it moves and leave an area quickly if conditions become unsafe. The one thing that a spotter must keep in mind at all times while mobile is safety. We've covered safety issues specifically in an earlier chapter, but they bear repeating here. Rules of safety must be adhered to religiously in a mobile environment. The weather presents hazards directly with rain, wind, hail, tornadoes, etc and indirectly with hazardous road conditions, distracted drivers and limited visibility. In the car we have several sources of potential distractions that we must stay aware of: a second spotter, radios, GPS, lights, cell phones, etc. Combine all of this with increased excitement levels and stress. Mobile storm spotting is not for everyone.

For those that do storm spot mobile, though, careful consideration must be paid to how we set up to observe severe weather. The first step is to find out where the storm is and the direction it is heading. Assuming it is a thunderstorm we will want to stay on the right rear flank of the storm. This gives us the best vantage point to observe severe weather if it develops and at the same time keeps us out of the storm. As the storm moves away from us we can change position to continue our observation.

When selecting a spot to observe from there are several factors to keep in mind. First choose a location that gives you as much view of the sky as possible. Try and avoid any obstructions such as trees, buildings, signs, etc on any side of you. Second have a planned escape route. Your escape route should be a roadway that is not likely to be made impassible by debris from the storm. Roadways with trees on either side or overhanging branches should be avoided altogether: stick with major roads such as highways. Third is to make sure you are still within communication range. Make sure you can still access the repeater and have cell phone reception. Fourth is to avoid being a traffic hazard. This goes back to safety. Do not set up on the shoulder of the road or in any way that may present a hazard to you or other motorists. Next, respect private property.

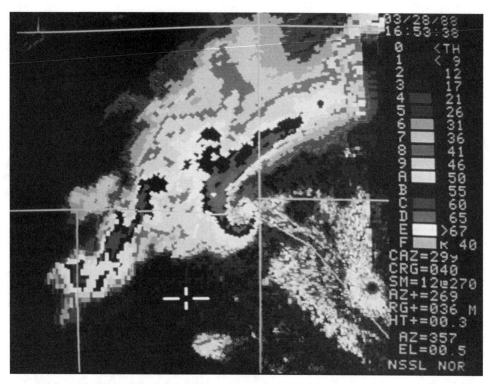

A radar 'hook echo', often a precursor of tornadic activity. (National Weather Service)

When you set up remember to do so from a location accessible to the public. What you do as a spotter is important but it does not give you license to be disrespectful to other's property. These steps to choosing a location for spotting are not merely suggestions but are all imperatives on the storm spotter that relate to storm spotting safety. Most of them can be addressed long before severe weather arrives. If you are dedicated to storm spotting mobile locate locations to spot from before severe weather strikes.

So now we have the actors in our storm spotter activation. The purpose of our response is to submit reports on severe weather. So what happens with these weather reports? Where do they go and how do they get used once they reach the NWS WFO?

Severe Weather Reports

Severe weather reports, they're the reason we serve as storm spotters. But if we were asked by John Q Public, "What happens when you call in a report?", how would we answer? In this section we will look at this and also discuss the importance of our storm reports, the role they play on the integrated warning system and the role they play in the larger severe weather team. We will then look at how the reports are actually received by the NWS. Finally we will look at what the NWS does once they receive them: preliminary *Local Storm Report (LSR), Post Event Reports (PER)* and *Storm Data*.

We know storm reports are important, but what do they actually provide? There are three things that storm reports provide to the NWS: ground truth information, situational awareness, assistance in verification efforts.

There are many attributes of a storm that the NWS can detect from their local field offices. Radar can be used to detect precipitation, tornadic signature (a *hook echo*), high winds (*bow echo*) and probability of hail. Satellites can be used to observe cloud structure and water vapor. All of these things are occurring far above the ground, far above the people and property that weather affects. What is missing is observation at ground level, information on what is being experienced. Ground-based weather stations can provide a lot of information but they cannot tell you if there is a wall cloud or the diameter of hail. Storm spotter help provide ground truth information on what is happening. This information, combined with data from radar, satellites, weather stations and other sources, helps give meteorologists at the NWS a more complete picture.

This more complete picture is what the second element of storm reports is all about: situational awareness. Information is what provides situational awareness. The more information received the more aware we are. We can get a feel for what this is by looking at a 911 operator as an example. The 911 operator receives a call for a crime in progress. They gather information from the caller (ground truth report). They can access previous calls for service at that residence or business (gather additional information). They dispatch police to the scene. Situational awareness is the ability to take all of the incoming information and paint a picture of what is happening to those that need to make critical decisions on course of action to take. For the forecasters at the local NWS office this kind of awareness helps them in issuing warnings, updating forecasts and providing critical life and property saving information to the public.

The information received from storm spotters also assist in verification efforts. As discussed above radar, satellites and weather stations provide valuable information but do not give the total picture. When information from these sources is combined with ground truth reports meteorologist can better verify what is happening and what might happen. Storm spotter reports also play a part in post storm verification efforts. Storm spotters may submit information after the event that helps NWS staff verify what happened, was the damage due to a microburst or a tornado?

The reports storm spotters submit also impact the *Integrated Warning System* (*IWS*). The IWS is basically the system in which weather information is gathered by the NWS, processed into a product, released to customers and an appropriate response is taken. The IWS has four elements: forecasts, detection, dissemination and public response. There are three primary users of data from the IWS: private meteorologists and media, emergency management and storm spotters and the general public. This system operates whether there is severe weather present or not. It covers the pre-event phase as well as the during-event phase.

Storm spotter reports positively affect several elements of the IWS. Reports coming in may be used to determine specific severe weather events or they may even be the catalyst that results in a severe weather warning. Amateur Radio storm spotters also play a key part in the dissemination element of the IWS. During severe weather we are often in communication with local emergency managers and NWS stations. We are relaying reports via radio on local conditions, severe weather events and other information of public interest. Many people in the general public listen to what we're doing over a scanner. This adds to the dissemination element. And our reports also assist in public response. The information that we provide can assist the NWS, media and local emergency management provide the public with necessary information on courses of action to take.

Now that we know the importance of our reports and the part they can play in the IWS let's take a look at how reports are received. Storm reports don't just happen on their own. Storm reports are received because of two factors: first are the relationships between members of the team and, second, the technical means that make it possible to submit the reports. If we think about the spotter training we received from the NWS we recall that there is an emphasis on teamwork and support. We, as storm spotters, are part of a team. Each member of the team plays a key part in the process of forecast, detection, dissemination and public response. When we activate as storm spotters we see this teamwork in action. We work with the local NWS office, emergency management and perhaps public safety. This team functions because of its ability to communicate amongst individual members and ultimately between members and the local NWS office. The support from the NWS helps keep the team running and effective. This support can be training (SKYWARN courses), answering questions and concerns from storm spotters, providing educational material and keeping lines of communication with spotters going, even when severe weather does not threaten. Storm reports are also aided by the relationships and collaborative efforts the NWS offices foster with local media and emergency management. We must remember storm reports may come from sources other than SKYWARN weather spotters. The storm reporting team consists of the NWS, local public safety officials, news media, storm spot-

The Importance of Being Honest

As storm spotters we are trained to report accurately. Trustworthy reports make volunteer storm spotters a valuable part of the warning system. As much as we try to relay the right information sometimes we may be mistaken. An isolated, honest error is tolerable, but routinely inaccurate or misleading information brings bad consequences, including loss of credibility and the risk of conflicting information delaying an important warning or advisory.

Some individuals deliberately have made false or misleading reports. In 2007, the NWS office in Milwaukee, Wisconsin, received over 25 false reports from the same computer in one weekend through the WFO's Web page for storm report submissions. The person, who appeared to have some knowledge of storm spotting and severe weather, falsely reported tornado damage and injuries. Local media interrupted broadcasts because of warnings being generated from the NWS office. Complicating the situation, legitimate severe weather was present in the area and a tornado warning had been issued based on storm spotter reports. The Federal Bureau of Investigation (FBI) investigated and since then, new measures have been put into place at the WFO to reduce false reports.

Unintentional false reports often result from inadequate training, poor communication and honest mistake, or a combination of factors. The consequences may be more training, better coordination, or perhaps restricting an individual from acting as a storm spotter.

Knowingly submitting a false storm report is against the law and is a violation of the False Statements Accountability Act of 1996 (18 USC §1001 – see the Appendix). Deliberately violating this federal statute can result in an individual being fined up to $250,000 and / or being sentenced up to five years in prison.

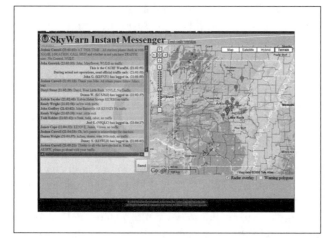

Instant messenger program allows storm spotters to relay storm reports to the National Weather Service. (Courtesy Central Arkansas UHF Group)

ters and emergency management.

As Amateur Radio operators we are well aware that there are many ways that messages can be relayed. Today we have more options than ever before: radio, cell phone, land line phone, Internet, etc. And of course there are communication systems developed by the NWS such as NAWAS. We have come a long way since the days of weather reports coming in via telegraph.

Without a doubt the Internet has proven to be a great tool for submitting weather reports. Over the last few years several different platforms have emerged as tools to submit storm reports. A basic and cost effective means has been the use of instant messenger (IM). IMs such as *Yahoo*, AOL and MSN of course are the first to come to mind. The NWS realized how valuable IMs can be and developed their own called *NWS Chat*. While not in use in all areas, it provides a secure means for NWS personnel, emergency management, SKYWARN and ARES members, members of the media and local, state and federal officials to communicate in real time and share critical weather information. NWS Chat is planned to be fully operational in 2010.

ESpotter, which we covered in a previous chapter, is another Web tool developed by the NWS. Like NWS Chat it is a secure system in that users must register for an account. However, it does not provide real time instant communication with NWS and other users.

Other Internet-based report tools are Webforms. Many NWS offices have an online form where storm reports may be made. The form asks for details that fit reportable criteria for that office. Unlike NWS Chat and ESpotter this is not a secured system. You do not have to be a registered user to submit a report. Typically, though, they will ask for identification and contact details to aid in verification.

Along with growth in Internet communications the NWS has also realized the value of mobile platforms. Mobile applications are available for cell phones and mobile Internet, so storm reports may be submitted from just about any place that can access these services.

The method of reporting that we're most familiar with, SKYWARN, is another way information reaches the local NWS office. SKYWARN reports may come from Amateur Radio SKYWARN and ARES groups or members of the general public that have taken SKYWARN training. Reports from the SKYWARN community may come in via a net control station, radioed direct to the NWS Amateur Radio station, or by some other means such as Webform or ESpotter. We can see SKYWARN as both a means and source of reports. Amateur Radio stations located at NWS office offer another means of communication while non-Amateur Radio SKYWARN spotters are a source that utilize other means.

Storm spotters know that what we report meets certain reportable criteria, eg a funnel cloud is reportable while lightning is not. There are times though that specific weather data is needed. The NWS has several methods for data to be reported automatically. These data can be a valuable part of the forecast and warning process.

One such weather data source is a *mesonet*. A mesonet is an automated weather station that reports mesoscale weather phenomena. Since the purpose is to observe mesoscale events, for example squall lines, mesonet stations are typically located close together and report fairly frequently, every one to fifteen minutes.

Two other automated reporting systems that may be utilized are the *Automated Surface Observing System (ASOS)* and the *Automated Weather Observing System (AWOS)*. ASOS stations, which are operated by the NWS, the Federal Aviation Administration and the Department of Defense, provide near real-time weather observation data to the NWS as well as the aviation, meteorological, hydrological and climatological research community. The information is updated each minute. Reports from ASOS stations typically occur each hour, but may be made more frequently if weather conditions exceed certain parameters. Data from ASOS stations include cloud height and amount to 12,000 feet, visibility to at least 10 statute miles, fog, haze, pressure, ambient and dew point temperature, wind direction, speed and character, precipitation accumulation and they may be set to look for changing conditions such as sudden changes in wind or pressure.

AWOS stations are operated primarily by the Federal Aviation Administration: they are not operated by the NWS or Department of Defense. These stations report weather data about every 20 minutes. The data reported are temperature and dew point, wind direction and speed, visibility, cloud coverage and ceiling and altimeter setting. Sensors may also be added that provide current weather conditions, detect freezing rain and provide lightning data.

Early on in the Cold War, NAWAS, the National Warning System, was implemented as a means of warning of enemy attack or a missile strike on the US. Since the Cold War ended NAWAS has been primarily used for natural and technical hazards. NAWAS is a telephone system, essentially a party line that connects government users and allows them to communicate with each other. The system has a couple of built in safeguards that help to keep it on line. It has built-in lightning protection and the lines avoid local telephone switches so they can stay on when circuits go down or are overloaded. Secondary users of the system include local emergency management, the NWS and local *public safety answering points (PSAP)*. These users can stay in easy contact using the NAWAS line. Information on severe weather may be relayed to the NWS and the NWS can use the system to disseminate severe weather information.

Another way the NWS can obtain storm reports is through direct solicitation. When the local NWS office puts out a statement containing weather information they may attach at the end of it a request for specific weather information and reports. For example, during a week where heavy rain is expected they may request storm spotters and local law enforcement to report any flooding. This typically may be found in issued hazardous weather outlooks. The NWS may also contact a local agency directly or make a request through a SKYWARN net for specific information.

Home weather stations also provide a source for storm reports. These weather stations can be used for reporting purposes in a couple of ways. If they can be connected to the Internet they can be set up to automatically report data to the *Citizen Weather Observer Program (CWOP)*. This information can then be accessed by the local NWS office, local public safety officials, emergency management and storm spotters. However, the station does not have to be connected to the Internet to be useful. Weather data may be monitored by the station owner and submitted to the local NWS office. For example, if the NWS office is seeking information on rainfall rates during a period of heavy precipitation the home weather station owner may monitor their station and make reports on hourly rainfall amount. And during hurricanes, home weather stations can be used to report weather data: barometric pressure, wind direction and speed and rainfall amounts.

Webcams also provide a way the NWS can receive severe weather reports. Webcams can serve many purposes. As we discussed in a previous chapter Weather Bug uses Webcams in their network of weather stations. And many TV stations have 'tower cams' that give a view of metropolitan areas. These can all be used by NWS personnel as a way of seeing what is happening. Like a storm spotter they can provide an extra set of eyes to observe weather conditions.

The aviation and marine communities are another source of severe weather reports. Both have vested interest in weather conditions and regularly monitor current conditions and trends. There are dedicated weather monitoring stations in each of these communities. We discussed earlier the ASOS and AWOS systems used in aviation. The marine community also has dedicated weather stations called weather buoys. These instruments collect weather data on the oceans of the world. Data collected includes: air and water temperature, wave height, barometric pressure, wind speed (sustained and gusts) and direction. Some buoys are moored while others drift. Data from these buoys can be monitored through several sources including: NOAA's National Data Buoy Center, Weather Underground and Storm Pulse. Also private citizens in the marine community can contribute valuable reports. Boaters around the world can relay detailed information on storm conditions.

We discussed earlier that the NWS does not endorse storm chasing. Obviously this does not prevent storm chasing from happening. Storm chasers can provide a valuable service to the study of severe weather. Many colleges, universities and research centers have active storm chase teams. And there are those that storm chase for personal or commercial interests. In either case storm chasers can provide local NWS offices with valuable storm reports. Storm chasers typically have significant meteorological knowledge and training. Many storm chasers also carry with them equipment to measure weather conditions, capture pictures and video and equipment to provide communications. All of this can make it possible for the storm chaser to quickly relay accurate, critical weather information to the NWS office.

Storm reports not only come in during the weather event, but also after the event. Following severe weather information continues to come in about the storm's aftermath. Videos and still images provide details on damage from the storm. Newspaper and television articles and stories provide details on damage and how populations were impacted. And accumulated data may be submitted, eg total storm rainfall in a particular area. As storm spotter we must be aware that post-event reports are critical. They help meteorologists understand more fully what happened, aid in verification efforts and make a full report on what happened possible.

Storm reports are absolutely essential in the IWS process. Because of this high level of importance it is necessary to use every possible source to obtain severe weather information. No source can be overlooked. Because of the goal of safeguarding life and property all reliable information must be considered.

So now that we've looked at the variety of ways that the NWS receives storm reports we have to ask ourselves, "What do they do with them?" The first step of course is to determine the accuracy of an incoming report. Generally each report is considered authentic until proven otherwise. Despite this benefit of the doubt accuracy of the report and consistency must still be established. Once this is done the report can move on to the next step, the preliminary *Local Storm Report (LSR)*.

The LSR is issued by the local NWS office. The LSR provides the *Storm Prediction Center (SPC)* which uses the LSR in hourly reports, adjacent Weather Forecast Offices (WFOs) and partners such as media, emergency management and storm spotters with information on hazardous weather events. The LSR gives another level of awareness of developing weather conditions.

The content of the LSR can include information on tornadoes, waterspouts, large hail, flooding, strong wind association with thunderstorms or marine gusts, or just about any other type of severe weather event. Like SKYWARN storm reports the LSR contains weather information that meets or exceeds warning criteria. The LSR content is limited to a single report of an event, not multiple reports of the same event. Likewise other information may be omitted such as unconfirmed events and events containing partial information that may result in confusion. Because of the information contained in the LSR they are issued as close to real time as possible. This means that the process of checking for accuracy and consistency must be done as quickly as possible.

LSR can also contain information beyond a single event. During the event an LSR may contain information that summarizes events that have occurred or are occurring. After the weather event an LSR may be issued which summarizes weather activities in the CWA. Each NWS WFO may issue LSRs however they see fit. This can be done as an LSR for each report during the event and a summary LSR afterward, some may compile a list of LSRs generated for each report received into a single report, others may issue only summary LSRs. Storm reports received in the post event period can help in compiling post event and/or summary LSRs.

Once the event is over the post-event reporting phase begins. In this phase additional information is gathered that will later help in compiling the storm data. During the weather event the reports coming in are much like information gathered at the scene of the crime, to use a police example. The information helps determine courses of action, in this case forecasts and warnings. In the post-event reporting phase, to continue the police analogy, it is much like investigators and detectives collecting additional information. This process helps fill in any blanks and verify events that did happen. Since one severe weather event can quickly follow another, time is of the essence. A second severe weather event can make this process difficult or even impossible.

So where do the post-event reports come from? We've already emphasized that storm spotters can assist in the process by submitting information on storm damage. The NWS WFO may make some follow up calls to verify events. Storm spotters may be asked, for example, if they observed any indications of tornadic damage. Public safety agencies may be called to see if they received any weather reports, since reports made to these agencies may not be immediately reported to the NWS due to other emergencies. Rural networks can also provide valuable post-storm information. Many rural areas may be more affected by communication interruptions during severe weather. The post-storm phase may be their first chance to submit reports. And NWS personnel may also gather newspaper clippings and video of storm events. All of this additional information may go into LSRs issued after the event. LSRs may be issued for up to seven days following an event.

Once the reporting and investigation is conducted at the local level the WFO begins the Storm Data (SD) process. The SD process takes information from local WFOs and presents it in a monthly publication that is distributed through the National Climatic Data Center (NCDC). The SD report covers "severe and unusual weather" that occurs across the US. The SD report contains highly detailed information about a storm. Local WFOs have 60 days from the end of the month that the event occurred to complete and submit SD.

SD reports are accessible by going to the National Climatic Data Center (NCDC) Web site[2]. A search tool allows users to find reports by state, date of event, county of event and event type. The user may enter additional search criteria such as tornado size, hail diameter, wind speed, number of injuries or deaths, or monetary damage to property or crops. Reports are then listed by location or county, date, time, type of event, magnitude, injuries, deaths and damage to property and crops. You may be surprised how many reports are available: a seven-month sample for Mississippi, January through July, contained 953 reports!

The reports that we make as storm spotters are part of a great flow of information about severe weather. Our reports and the reports from other sources are essential in the IWS. The products that are generated from the reports are the *Local Storm Reports*, which provide valuable information to storm spotters and other NWS partners and *Storm Data* which provides detailed information about severe weather events. Many of us, as storm spotters, have seen and perhaps used information contained in LSRs. And at times a report we may have submitted may have caused an LSR to be issued. But our reports have a life span far beyond the LSR. They are a part of the process of keeping an archive on severe weather and contribute to a better understanding about weather.

ARES AND SKYWARN

Several times we've mentioned the dual response nature of ARES and SKYWARN. Let's take some time to discuss the relationship between the two organizations.

ARES, the Amateur Radio Emergency Service, is an organization of licensed Amateur Radio operators that voluntarily register the qualifications and equipment for communications duty in the public service when disaster strikes. The only requirements for membership are an Amateur Radio license and a desire to serve. Coordination of ARES falls under the ARRL and is supervised by the Field and Education Resources Manager. The organizational divisions of ARES are national, section, district and local. These levels are over-

Damage from Worcester, Massachusetts ice storm, December 2008. (Rob Macedo, KD1CY, photo)

seen by the Field and Education Resources Manager, Section Emergency Coordinator, District Emergency Coordinator and Emergency Coordinator respectively. ARES members may respond to disasters that can include anything from hurricanes to wildfires, search and rescue activities to power outages, tornadoes to hazardous material spills. ARES is basically an Amateur Radio all-hazards response mechanism.

SKYWARN is a separate program from ARES. It is run by the National Weather Service and is coordinated by a meteorologist at the local NWS office. The local WFO SKYWARN coordinator generally oversees the local SKYWARN program in their CWA, sets up SKYWARN classes and may recruit some local coordinators to provide assistance. They also work with the Amateur Radio operators that run the Amateur Radio station at the local WFO. Unlike ARES, SKYWARN is not an all-hazards response mechanism: SKYWARN is limited to severe weather.

Many Amateur Radio operators are involved in one or both organizations. The Amateur Radio emergency response, regardless of disaster type, is handled through ARES. It is the responsibility of the Emergency Coordinator to oversee the local Amateur Radio response during a disaster. This includes severe weather. Because of this, in many places ARES and SKYWARN are often synonymous. When severe weather strikes we go into an ARES / SKYWARN mode. Severe weather reports are handled through the SKYWARN system, while emergency communication assistance is handled through ARES. This would be the dual-response aspect of ARES and SKYWARN.

Dual response of ARES and SKYWARN includes some assumptions. First is that both groups must exist within a local area. Second that dual response indicates shared membership, ie that ARES members are SKYWARN members. Third is that the relationship between the two is understood. Fourth is that individuals that are members of both organizations have been through the appropriate training for both organizations.

There are also times where we go in to a single response mode. For example a weather event that does not require emergency communication assistance. Many times we are called on for SKYWARN duty and nothing more. Other times there exists an emergency communications need but no severe weather event that would require a SKYWARN response. A hazardous materials spill would be an example. While information on such an event might have to be relayed to the NWS it does not meet the requirements for SKYWARN activation.

Membership in SKYWARN does not automatically include ARES membership and *vice versa*. SKYWARN requires very specific training for its members. And ARES includes training that SKYWARN does not, such as NIMS (National Incident Management System) / ICS and EM-COMM classes.

Working with Volunteers

We all realize that SKYWARN and ARES are volunteer organizations. Volunteerism is the heart of being an Amateur Radio operator, particularly during times of emergency. We've discussed how we are organized during disasters either through SKYWARN or ARES. For those in volunteer leadership positions, such as local SKYWARN coordinators and ECs, it is valuable to have some understanding of how volunteer organizations work, their strengths and weaknesses and how to handle problems that may come up with volunteers. We can learn a lot about these things from the emergency management community.

One of the first steps that we can take is in training. FEMA offers a course through their independent study division called 'Developing and Managing Volunteers'. While this course is intended for those in the emergency management field it offers a great deal of insight about emergency volunteers for us as well. Local SKYWARN coordinators and ECs can learn a great deal about volunteer management, recruiting and handling stress among volunteers.

The first step for the SKYWARN coordinator or EC is to be familiar with the relationship we have with the served agencies. In the case of storm spotting this will primarily be the NWS and emergency management. While volunteers provide served agencies with many benefits there are also challenges in working with volunteers. These challenges may be real or perceived. We're pretty familiar with the benefits that volunteers provide: cost effective services, access to a broad range of expertise and experience, free up paid staff to focus on other tasks, serve as a link to the community. So what are some of the challenges? Training volunteers takes time. Can you say you were ready for all types of severe weather after one SKYWARN training class? Volunteers are not permanent. There are some who feel that technically competent people do not volunteer or believe that volunteers lower professional standards. There are insurance and liability issues with volunteers. And there can be a feeling by paid staff that volunteers are competing with them. While there are some things we cannot change, there are some things that we can affect. As a local storm spotter coordinator you can play an important role in promoting benefits and alleviating challenges. This is not done alone: remember we're in a relationship with a served agency. Working with the staff of your local NWS WFO and local emergency management office is important in developing a strong volunteer-served agency partnership.

Another issue in volunteer organizations is recruitment. SKYWARN and ARES both have very basic membership requirements. SKYWARN requires completion of a SKYWARN class offered by the local NWS office and the ability to communicate reports to the local office. ARES requires an Amateur Radio license and a desire to serve during times of disaster. Does this mean that both organizations only draw those that are highly qualified, competent, dedicated emergency communicators that can handle the stress and demands found during times of disaster? No. Is there a mechanism in place to weed out volunteers that pose a liability to the organization? The only mechanism in place is found in federal regulations. It is a crime to submit false weather reports to the NWS (see the Appendix) and as Amateur Radio operators we have Part 97 of the FCC rules. Does this pose a serious challenge to the local SKYWARN coordinator or EC? Yes. So how can a coordinator or EC develop the best local emergency communications group possible? Through these steps: recruit, train, evaluate and recognize / retain.

SKYWARN and ARES both recruit members through a number of sources: Websites, classes, pamphlets, etc. At the local level the SKYWARN coordinator or EC can develop methods to recruit Amateur Radio operators to both of these organizations. The first step is to develop a way of presenting information to potential volunteers that explain what the expectations are. For our purposes we are going to focus on severe weather and storm spotter response. So we will want to explain that there is a training requirement, the NWS basic SKYWARN class. We will explain the role we play, as storm spotters we provide reports on severe weather conditions and as Amateur Radio operators we support NWS communications through our net control stations. We need to explain that severe weather response is not for everyone. Severe weather can make for high stress and high risk environments. We need storm spotters that are safe, responsible, reliable, trustworthy and can handle a high stress situation. We also need to explain what equipment an Amateur Radio storm spotter needs. And we need to explain the relationship Amateur Radio has with the NWS and local emergency management. People are more likely to volunteer if they are presented with as much information as possible on what is expected of them. Once we have a plan in place on how to present the information we need to look at where we can find volunteers. Consider several possibilities: hamfests, club meetings, VE sessions, SKYWARN classes and nets.

Once we have volunteers we need to move to the next step, training. Obviously the first step for a storm spotter is SKYWARN training. We don't want to stop there, though. We need to emphasize other training possibilities: EMCOMM classes, Advanced SKYWARN, FEMA courses. Refer back to the training chapter for ideas. We cannot force people to commit to training beyond what is required. Make every effort, though, to encourage people to take advantage of every training opportunity. Work with your local NWS office to bring regular SKYWARN classes to your area. Just because you've taken it once doesn't mean that you can't take it again. Have local SKYWARN meetings where information about severe weather can be presented. Have veteran SKYWARN storm spotters as guest speakers. Be creative and encourage a learning environment. It is very easy for us to get lax in this area when severe weather is not threatening or during seasons of regularly good weather.

Evaluation is another key in a good local volunteer program. Evaluation occurs after an event. After a SKYWARN activation has ended and the final assessments are being done evaluation of individual performance is needed. This is what tells us if there is a problem, a training issue, or if a member of our group went above and beyond.

And finally is recognize and retain. If we don't recognize volunteers for their effort we will be surprised how fast they leave the group. Human beings by their very nature expect some form of reward for their efforts. Volunteers are no different. We may not receive monetary compensation for what we do, but payment in the form of recognition can be just as valuable. The amount of time it takes to recognize a job well done and make that recognition known is negligible. Recognition can come from the local level from the local coordinator or EC. It can come from a district or section level, DEC / SEC / SM. The ARRL has a certificate that can be used for recognizing the efforts of an Amateur Radio operator (Certificate of Merit, found under Field Services Forms).

When working with volunteers we also have to learn how to handle volunteer stress. Whenever we volunteer as Amateur Radio operators to assist during times of disaster stress should be anticipated. Severe weather and storm spotting are no different. There are ways though we can address and handle stress amongst our volunteers.

Before severe weather ever strikes we can address the issue of stress through training. There are classes available that teach stress management skills. These may not be available everywhere or may cost money. An approach we can take is by utilizing expertise found in our local public safety agencies. Invite to a storm spotter meeting a police officer, fire fighter, medic, or 911 dispatcher and have them talk about how they handle stress on the job.

During a severe weather event ECs and SKYWARN coordinators should make sure volunteers are well matched to their task. If someone cannot handle a fast paced, high stress environment then placing them in the emergency operations center may not be a good idea. Also make sure that volunteers get regular meals and breaks and that they are rotated out after a reasonable length shift. This is particularly for severe weather events of long duration such as floods, ice storms, or hurricanes. These are the events that most likely will be SKYWARN and ARES events.

After the storm make sure you talk with the volunteers and find out what kind of stress they were under. Was there some way that the stress could have been alleviated? Do they need additional help such as Critical Incident Stress Debriefing? Severe weather can result high levels of stress and trauma. Imagine if a storm spotter that was operating mobile came home to find that their neighborhood was hit by a tornado. Storm spotters are not immune to the trauma of severe weather destruction.

The ARRL Emergency Communications Handbook offers some additional tips to dealing with stress and stressful situations:

- Delegate some of your responsibilities to others. Take on those tasks that you can handle.
- Prioritize your actions, the most important and time sensitive ones come first.
- Do not take comments personally. Mentally translate personal attacks into constructive criticism and a signal that there may be an important need that is being overlooked.

Storm Spotting & Amateur Radio

- Take a few deep breaths and relax. Do this often, especially if you feel stress is increasing. Gather your thoughts and move on.
- Watch out for your own needs: food, rest, water, medical attention.
- Do not insist on working more that your assigned shift if others can take over. Get rest when you can so that you will be ready to handle your job more effectively later on.
- Take a moment to think before responding to a stress causing challenge. If needed, tell them you will be back to them in a few minutes.
- If you are losing control of a situation, bring someone else in to assist or notify a superior. Do not let a problem get out of hand before asking for help.
- Keep an eye on other team members and help them reduce stress where possible.

Imagine a SKYWARN activation for a severe thunderstorm. The NWS has indicated that there is potential for strong winds, large hail and possibly a tornado. We activate our SKYWARN net. There is a net control station, a relay station, a dozen or so home-based spotters and about half a dozen mobile spotters. Our area of responsibility is County X. Net control is relaying information to the storm spotters on what the radar is showing and where safe areas would be to set up and observe. During the net a mobile spotter disregards net controls instructions and takes on what can only be described as a 'storm chaser' approach with the severe weather. He states "I know what I'm doing, don't tell me what to do!" and follows the storm on its path into County Y. He has become a 'loose cannon'. His behavior is dangerous to himself and makes the organization as a whole look bad. The next day, after the storm had passed, he admitted his wrong doing and the problems it caused. What fed this kind of behavior? Simply put, *ego*. When working with volunteers stress is just one challenge that we face. We are also challenged by egos. All of us have an ego, some big, some not so big. During a stressful situation, such as severe weather response, egos can come to the surface. There may be control issues, 'loose cannon' personalities, tempers and all sorts of ugly behavior. These can create some embarrassing and at times dangerous situations. As coordinator or EC you must be ready and able to deal with these situations.

Damage from the May 2007 EF-5 tornado in Greensburg, Kansas. (Keith Kaiser, WA0TJT, photo)

During the severe weather event it is best not to confront ego issues head on. Address the specifics after the event is over. A skillful coordinator or EC will diplomatically and tactfully rein egos back in. It is better to rein in the loose cannon and keep them as a productive storm spotter than to anger them and encourage further behavior.

If you are in a position of managing any of your local storm spotters you have a challenging job. It is often thankless and carries with it a significant level of stress. Prepare yourself by training and learning how volunteer organizations work.

AFTER THE STORM

Once the storm has passed our role as Amateur Radio operators and SKYWARN storm spotters is not over: it enters a post-event phase. In the post-event phase we have four basic tasks. First is to continue our support of NWS efforts and contribute storm damage reports and any other reports that may have not been submitted during the event. Second is to submit appropriate reports to the EC, DEC, SEC and ARRL on Amateur Radio activities in support of the NWS and severe weather response. Third is to conduct an after-action report detailing our activities and response. Finally, fourth, to take part in any necessary debriefing that may occur. Let's look at each of these.

Damage Reports

At several points we've mentioned the importance of storm damage reports to the NWS. Following an event meteorologists from the local WFO will set out to photograph and document damage related to the storm. Typically, though, local NWS offices do not have large staffs that can fully handle this task. This is where SKYWARN trained storm spotters can assist. Weather affects where we live. Storm spotters are familiar with their communities. We know what areas are accessible and inaccessible following a storm. And we can quickly spot damage from the storm. By utilizing the SKYWARN storm spotter network the local NWS office gains extra hands for the task of documenting how a severe weather event impacted our area. This information can be used in developing the local storm report, determining the nature of the event, determining the severity of the event and for future planning.

Post-storm reports are not limited to damage. Some spotters may not be able to relay in storm reports as events happen. Weather reports may be submitted after the event has ended. This information is also valuable. And some data such as total rainfall may not be fully known until after the event has ended.

ARRL Report Forms

In some areas the local EC may serve as the local Amateur Radio SKYWARN representative too. This may be in an informal, unofficial, capacity or it may be by appointment from the local NWS WFO. In other areas the Amateur Radio SKYWARN representative is a separate position from the EC. In either case when Amateur Radio SKYWARN members are activated for severe weather it is a matter of concern for the local EC or, if there is no local EC, the DEC. For our purposes we are going to assume that the EC and Amateur Radio SKYWARN representative are two separate positions, held by two separate individuals. Severe weather response and SKYWARN activity can fall into two areas that Amateur Radio operators specialize in, public service and emergency communications. Because of this we will want to document our SKYWARN activity for our EC / DEC. Beyond the local level Amateur Radio severe weather response also concerns our section leadership, the SEC and SM and ultimately the ARRL.

The way we report our activities during severe weather are by using the standard ARRL Field Services Division (FSD) report forms. These are found on the ARRL's Web site on the Amateur Radio Public Service page[3], listed as 'FSD forms'. There are several report forms that pertain to severe weather response; the specific forms we are going to cover in this section are: *FSD-157 Public Service Activity Report, FSD-212 Monthly DEC / EC Report, Form C EC Annual Report* and *FSD-96 Monthly Section Emergency Coordinator Report to ARRL Headquarters*. All of these forms may be found in Appendix 5 at the end of this book.

So let's start at the local level. The first report form we will cover is the *Public Service Activity Report*. We can

ARRL *Public Service Activity Report* form *FSD-157* (see Appendix 5 of this book).

consider SKYWARN activation as a public service activity because its very purpose is to provide information to the NWS in support of their mission to help safeguard life and property. This is an activity we want to be certain to document. By using the ARRL form *FSD-157* we do this and our report can be used to show to Congress, the FCC and public officials exactly how valuable of a service Amateur Radio provides. The *FSD-157* form allows us to document key data about our activation. First is the nature of our activity. SKYWARN activation would best fall under the category of Communications Emergency. While we are not likely going to be replacing normal communications during SKYWARN activation we are supplementing normal communications by providing additional means to relay information to the NWS. Next we document the nature of our activity and the area involved. We will want to be more specific than just saying "SKYWARN activation". This is being done in the post storm period so we should have a better idea on event specifics. Instead indicating "SKYWARN activation for F-3 tornado" would provide better details. Indicate the area involved by city or county. Remember that people may be reading your reports that have no idea where "the old saw mill by the lake" might be. Once we have our activity and location indicated we can specify the number of Amateur Radio operators involved and date / time of event. Remember to include mobile and home-based spotters, relay stations, net control stations and Amateur Radio operators that checked into the net that may have only given an incidental report on the weather: when indicating date and time be sure to specify if times are local or UTC (GMT). We're also going to want to be sure to indicate the total "person-hours" for the event. This is the amount of time each individual was involved in the activation. For example, let's say K5DSG was involved for four hours, KD5JHE was active for two hours, KC0IKK was active for one hour and KC5UDM was active for three hours, so our total 'person-hours' would be 10. This figure is then used to calculate "person-power" cost, which is currently $19 per hour. We will then combine this with the estimated cost of equipment used to determine a total estimated cost of service. So if we continue with our example we can determine the "person-power" cost for our four SKYWARN members is $190. We combine this with the total cost of equipment used: two mobile stations, two home-based stations and a repeater for a total of $4480. This means that our total estimated cost of service is $4670. Now, to document all of this we must keep careful records during SKYWARN activation and possibly do some follow up once the activation is over. Net control, or some other appointed station, should keep a log of who was involved in the activation and the amount of time they were active. Equipment costs can be checked on after activation ends or can be estimated. Individuals participating in SKYWARN activation can help be keeping a record of their activity time.

Now that we have the basic data of the way we can go on to provide information on nets, traffic, served agencies and other additional data. We are going to want to include in our report information on the nets and frequencies that we used during activation. When we list the net saying "Oxford-Lafayette County SKYWARN Net" is not enough. We want to include the repeater information along with the net name. This provides documentation on the types of frequencies that are utilized for public service. So if during activation we used our local SKYWARN net on the 2 meter repeater and relayed information to the NWS via an HF net on 80 meters, we will want to indicate both on the report. Next we will list any messages passed during the activation. We can include here storm reports sent via Amateur Radio to the NWS. Now we are going to document the agencies that we supported during activation. Obviously we are going to indicate the NWS. Be sure to specify which office was supported. We will also include any local agencies that we may have assisted during activation. This could include local emergency management, public safety, or government officials. Next we document the specific call signs of Amateur Radio operators that were major participants. Net control logs should provide this information. And before we get to the reporting party's information we should take some time to provide some additional comments if necessary and photos, newspaper clippings, or other data if available. Remember, this is a report, the more information the better.

Finally, we are going to provide information on the reporting party. There are no hard and fast rules on who should submit this report. Ideally it would be a designated, local amateur who was involved in the activation and the documentation of activities such as the local EC / DEC, net manager, or SKYWARN coordinator, but any Amateur Radio operator that has knowledge of the event can submit the report. Once all of the documentation is complete we will send the completed for to the ARRL. This form can be filled out online or can be printed out and mailed in. If the form is completed by someone other than the local EC / DEC a copy should be sent to them as well. While there is no set time frame for the report to be submitted it is best to do so as soon as possible after the event. The longer it is put off the more likely it will be that information will be omitted or less accurate.

The ARRL Field Organization provides emergency communications management at local, district and section levels with the Emergency Coordinator, District Emergency Coordinator and Section Emergency Coordinator respectively. All of these positions fall under the general leadership of the Section Manager. When it comes to Amateur Radio severe weather response each of these have an interest.

The EC is appointed by the SM and works along with the DEC and SEC as well as Official Emergency Stations. The jurisdiction of an EC is typically a city or county. The EC "prepares for and engages in management of communication needs in disasters". While the EC is an ARES function they are part of severe weather response. This is largely due to the dual response role of ARES and SKYWARN. There are several specific duties of the EC which are listed on the ARRL Web site and in *The ARRL Emergency Communications Handbook*, we won't list them all here but will discuss duties that pertain to severe weather response.

An Amateur Radio SKYWARN Report Form?

In this chapter we discuss forms designed for specific local events (*FSD-157*), EC / DEC reports (*FSD-212* and *Form C*) and for SEC reports (*FSD-96*), but is there a form specifically for SKYWARN activations? Currently no, there is no such form. Is it possible to develop one? What benefits would it have?

It is entirely possible that an Amateur Radio SKYWARN activation form could be developed. Severe weather response plays a large part of our emergency communications and public service involvement. And severe weather response is seldom like other emergencies such as hazardous materials incidents, wildfires, parades and festivals, search and rescue and any of the other incidents that amateurs assist with. When we respond to severe weather we generally do so under the SKYWARN banner. Our SKYWARN response comes from the relationship between the ARRL and the NWS. We respond to gather specific weather information that is relayed to the NWS. We do not go out and radio in "it's raining" or "nothing to report". We are trained to look for specific events. So if we, the Amateur Radio community involved in SKYWARN represented by the ARRL and the National Weather Service, were to develop a SKYWARN report form what would it include? How would it benefit SKYWARN? Who would make the report? What timeframe would it be submitted in?

A SKYWARN report form would focus on the activation of local Amateur Radio operators that are SKYWARN trained that actively participate in a local SKYWARN activation. The form, similar to the *FSD-157 Public Service Activity Report*, would include information on the reason for activation, eg winter storm, thunderstorm, heavy participation. Further details could be added as necessary. Date, time and duration of activation would also be included. Borrowing from *FSD-157* the form could also include person-hours, person-power cost, equipment cost and total estimated cost of service. The amount of money saved, or value of services provided, is often strong factors in determining the overall value of a service. And taking again from *FSD-157* we would also want to indicate net information and call signs of major participants. The key differences between this form and *FSD-157* would be the inclusion of specific weather reports relayed. Since most SKYWARN groups report on standard criteria: funnel cloud, tornado, wind speeds in excess of a set minimum, hail above a minimum diameter, etc, the form can indicate the number of reports called in for each event. The method of reporting can also be indicated: report made via Amateur Radio net, Internet form, Spotter Network, etc. Finally the form would indicate who submitted it.

The content of our Amateur Radio SKYWARN report form can be debated. Some groups may want to include more, some less. Ideally the form would be the product of a collaborative effort between local amateurs and their respective NWS WFO. The form could have several potential benefits. It would provide information about the amount of Amateur Radio activity in SKYWARN in a local area. It would provide details on the value of Amateur Radio to the NWS. It would provide useful information on the number of reports about specific event that amateurs are calling in. For example, we may be able to see that certain weather conditions that are reportable are being overlooked, such as wind speed. It would also give an idea on what methods are being employed to report severe weather to the NWS. The overall benefit is that, assuming the final report goes to the ARRL and the NWS, both entities are receiving the same information on local activity. This can be useful in identifying training needs, recognizing potential local SKYWARN leaders / mentors and gathering data at a local level that can be used to develop a picture of SKYWARN activity within the NWS CWA. It would also encourage the development of SKYWARN coordinators that assist the NWS with local SKYWARN activity.

Such a report would have to be submitted on a timeline, though. If we use the LSR and Storm Data reports as guides we could possibly set a submission timeframe of within a week of the event. This kind of detail would have to be agreed on and established by all parties.

OK, so what reasons are there to *not* have such a report form? First is that while it does provide great information about the Amateur Radio SKYWARN response it misses large sectors of SKYWARN storm spotters. We must keep in mind that storm spotters come from many different areas: emergency management, public safety, education, business, local government. Amateur Radio operators are only a part of the SKYWARN community. Also it could be argued that the *FSD-157* is adequate in reporting severe weather response from Amateur Radio operators. And since such a form would have to be somewhat customized to meet local needs (weather is not the same everywhere) it would be difficult to assemble all the data collected from the forms.

The idea for an Amateur Radio SKYWARN form is brought up here strictly for discussion and to get us thinking. Amateur Radio has, since SKYWARN's beginning, been a unique asset and has offered some great advances in the way severe weather reports reach the NWS. The reason to report what we do is so that we can learn and grow from our experiences. This has been done successfully in the ARRL Field Services for many years. Is it possible that a similar reporting system benefit the relationship between Amateur Radio and the National Weather Service? That is open for debate.

The EC's duties include a variety of tasks. They are the point of contact for amateurs interested in getting involved in emergency communications. They help develop and provide training for amateurs. They develop, and maintain, working relationships between the local Amateur Radio community and the served agencies at the federal, state and local levels. They provide input in the development of local emergency plans. They establish local emergency communications networks and regularly test them with exercises and drills. They serve as the emergency communications point of contact during a disaster. And they work to grow the ARES program. Basically, your local EC is your local Amateur Radio emergency communications manager. They can provide a great resource for local SKYWARN activity by assisting with training, exercises, information and being the point of contact between SKYWARN storm spotter activity and local officials.

There is one important duty of the EC that relates very directly to SKYWARN and storm spotter activity: "Report regularly to the SEC, as required". In the absence of an SEC the EC reports to the SM. There are two reports that the EC submits to the SEC / SM, the *Monthly DEC / EC Report, FSD-212,* and the *EC Annual Report, Form C*. On the monthly report the EC reports the number of public service events and emergency operations conducted within their jurisdiction for the month. SKYWARN training classes that are sponsored by local Amateur Radio SKYWARN teams would constitute a public service event as long as the general public was invited to attend. SKYWARN activations would constitute an emergency operation, whether or not they are associated with and ARES activation. Besides the number of public service events and emergency operations the EC also reports the number of 'person-hours'. On the annual report the EC reports on the agencies that are served by the ARES group. Again, going back to the dual response nature of ARES and SKYWARN, the EC may want to include the NWS as a served agency through SKYWARN. If an area has someone managing the Amateur Radio SKYWARN team other than the EC they should forward SKYWARN activation information to the EC to be included in monthly and yearly reports.

Just like the function of the EC as a local emergency communications manager, the DEC serves in a similar capacity at an area level. The DEC's jurisdiction may be several counties or an area within a section. A key role for the DEC in regards to SKYWARN is that they serve as a local EC for counties or other jurisdictions that do not have one. If your area lacks an EC you may want to find out if you have a DEC to turn to for emergency communications guidance. As far as reporting goes, the DEC makes similar reports as the EC and submits them to the SEC.

The SEC is the Amateur Radio emergency communications coordinator at the section level. Like ECs and DECs they have a duty to submit reports on emergency communications activity within their section. The form that is used, *FSD-96*, is similar to the one used by ECs and DECs. It includes information on public service events and emergency operations conducted each month. The SEC reports this data directly to the ARRL.

AFTER ACTION REPORTS

'Lessons learned' has been a buzz-word in the public safety for several years now. Lessons learned from 9/11 and Hurricane Katrina are still being discussed and as we experience other critical events new lessons learned are added. Whether we respond to severe weather as SKYWARN spotters, ARES members, or as part of some other group a valuable part of the activation process is assessing our performance, determining what went right or wrong and taking corrective action. Through this process we can identify objectives of the activation and determine to what degree they were met, the *After Action Report (AAR)* makes all this possible.

Writing AARs are standard in military, public safety and emergency management. Any time there is a disaster, major public event, or critical incident an AAR will be written to assess the response. Disaster exercises and other training events also would warrant an AAR to assess response and determine if the exercise objectives were met. In this section we will go over what an After-Action Report is, who writes and reads them, when they should be done, why it is done and how it is done. We will also look at the differences in After Action Reports written for critical incidents and exercises. As we go we will see how AARs can be used in Amateur Radio response to an emergency. And finally we will go over how to get your group started in writing an After Action Report.

What is an After Action Report?

The US Army provides a great explanation on what an After Action Report is. In *A Leader's Guide to After Action Reports* they identify four key aspects to an After Action Report[4]. First, it is a discussion of an event. While the Army may be concerned with an event such as combat and event could be a hazardous materials incident, earthquake or tornado. Second, it focuses on performance standards. Very

Photos of storm damage in After Action Reports can help the National Weather Service understand the extent of damage and what caused it. Tornado damage in Hoisington, Kansas, April 21, 2001. (Robert Haneke, WG0Q, photo)

> **After Action Reports: Why are they so Important?**
> The After Action Report (AAR) allows us to identify strengths and weaknesses in our response. When we identify weaknesses we can identify ways to improve our response in future emergencies. And over time AARs will make it possible to see trends and patterns in our response (both good and bad). AARs also allow those involved in the response to see the big picture of the event. They may see what other responders have done and discover solutions to problems or provide solutions to other responders.
> There is another advantage to writing AARs. They can provide resources for responders outside of the affected area. By making them available to other agencies and groups it is possible to expand the knowledge base on response to emergencies. So going back to our ice storm example, when the final AAR is compiled it can be made available to ARES / SKYWARN groups and served agencies, such as the National Weather Service or Red Cross, outside the disaster area.

seldom do Amateur Radio operators respond to an emergency situation without standards. In fact we pride ourselves and are often commended on our ability to excel in providing communications during a crisis. Third, an AAR explains to the reader what happened and why it happened. And finally the After Action Report provides insight to sustain strengths and improve on weaknesses. We are not perfect and mistakes will happen. We only fail at what we do when we do not learn from our mistakes and improve. By using a tool such as the AAR we can welcome mistakes in our response since it gives us a way of building from them. There may be differences on what an After Action Reports contains depending on who wrote it and what event it covered but these basic goals of the report remain the same.

Who Writes the AAR and Who Reads it?

The AAR may be a collaborative effort. Early on in any critical incident someone should be designated as the person responsible for compiling the After Action Report. This person should be familiar with planning the response, the organizations involved and the objectives of the response. Others may contribute to the report. Individuals involved in responding to the emergency can contribute by submitting reports on what happened, how they responded, problems encountered and objectives met. In the end these account may be compiled and put into the final report either in their entirety or incorporated into a summary of reports received. But for these reports to be accurate there must be a process in place to document events as they occur. Some ways of doing this are: logs and journals, written messages, action plans for specific events and public information and media reports.

This process can be used in writing an Amateur Radio After Action Report. Let's use a major ice storm as an example. Such an event can cover a wide area, affecting several states. In this case the final AAR may be a collaborative effort between the Section Emergency Coordinators in the affected area, each one putting together an AAR for their area with the final AAR being a compilation of each section's AAR. Of course to do this requires collecting data from the local responders. This can come from reports written by local Emergency Coordinators, District Emergency Coordinators, SKYWARN storm spotters, SATERN members and any other groups that may have responded. And just like any other event that may warrant an AAR there must be some form of documenting events as they occur. Amateurs are familiar with logs and can use these to keep track of reported events, messages and net activity. Field responders may keep a notebook journal to keep track of events and activities. And software is available to help keep track of resources and times. Needless to say, this can be a massive task. But the rewards of a good AAR are worth this critical follow-up work after the event. Later we will look at ways of streamlining the process of collecting this data.

So once it does get written, who reads it? An AAR is designed to be a useful tool to those that plan response, respond and assist in future planning. The AAR would definitely be of use to all parties involved in the emergency response. This could be the individual storm spotter, net control operators, local and section leaders and any number of served agencies that may have been involved. We can't look at this as a report isolated to Amateur Radio, but as a report of one component of a larger response. Sharing these findings with other responders can benefit the Amateur Radio response as well as their response. What an AAR is *not* is a press release or any form of public information statement. It may contain information that isn't really suited for public release. This is not necessarily because there is anything to hide, but scrutiny by readers not familiar with all aspects of the emergency or response may not be entirely beneficial. If a public information statement is needed that is a separate matter handled by a public information officer or spokesperson.

When Should the AAR be Written?

Ideally the AAR should be written as soon after the event as possible. Depending on the size of the event this may be difficult to do. Documentation from the event needs to be gathered, responders or leaders may need to be interviewed, surveys may need to be conducted and perhaps workshops made up of responders may need to be held. A lot of work goes into gathering the information needed for a good AAR and it must be done in a timely manner. By waiting too long

Cover sheet of the extensive After Action Report written following the 'Hurricane Joshua' exercise in 2007.

after the event we may forget some details, neglect documentation, or not provide enough detail in the report.

An Amateur Radio AAR is no different. Data must still be collected as soon as possible after the event, analyzed and put into report form. During Hurricane Gustav an EOC staff member told me that after every hurricane they talk about getting everyone together to discuss the event, problems encountered and write a formal AAR. But each time, they told me, everyone is so worn out from the demands of the emergency that little is done and a far less formal AAR is eventually compiled. Events such as hurricanes and ice storms and any other emergency event can take huge tolls on those that responded. But the event is not over when the threat is gone. The event is over when we have included in our recovery effort a plan for better responding to the next emergency.

What Goes in the AAR?

AARs come in different types. AARs for large scale events such as hurricanes and earthquakes can contain massive amounts of information while events such as an isolated tornado or hazardous materials spill may contain less. And the goals of an AAR for a disaster exercise may be different than for a real emergency, since the purpose of the exercise is training. Regardless of the size of the event or whether the AAR is formal or informal the AAR contains certain key information.

The AAR begins with introductory information. This may include the type and location of the event. It should contain maps of affected areas (if available), a timeline of events, date / time of any proclamations or declarations (ie state of emergency, federal disaster area, etc) and duration of event. Also included would be a general description of the event.

The second part covers a discussion of response. This discussion will look at different levels in the response effort. It will likely cover field level response, local government response, interactions within the operational area and the regional area, interaction with state level agencies and interaction with federal agencies. It will focus on planning, logistics, finance / administration and multi / inter-agency coordination (this should be familiar from NIMS certification). Granted, this may be beyond the scope of an Amateur Radio response. But this is a section that can help us focus on our response as part of a coordinated effort at local, section, division and regional levels. There are times, though, that this may not be needed, such as when the event is isolated and does not involve participants at anything higher than a local or state level.

Next the AAR covers the participants and systems involved in response. This is where we cover the individual agencies and groups involved. We may first start off by addressing mutual aid systems that were involved. During a hurricane in the Gulf Coast area a memorandum of understanding may be utilized to help coordinate efforts between affected sections in the Delta Division. This section will also address the participants in the response. Most times Amateur Radio is one part of a large-scale response. Other players may include public utilities, Red Cross, National Weather Service, Salvation Army, the media and public safety. There may also be interaction with other levels of response. Assessment of these interactions is important. When Amateur Radio responds we do not go alone. We generally assist other agencies and groups and an assessment of our interaction with them is critical. And, as you can probably guess, they are assessing their interaction with us.

Next the AAR will address training needs. This is where we look for areas in our response that can be improved on by further training. There is no such thing as a perfect response. If we get into the mindset that our actions during a disaster have no room for improvement then we are setting ourselves up for an even bigger disaster in the future. Talk with all involved in the response and get feedback on what went wrong as well as what went right. Problems may be small or large but either way should be addressed so they don't come back to haunt us in the future. This is also a good spot to focus on some training that the group may need. Does your group have problems with radioing in weather information that is not reportable? Identify that issue here and discuss ways that the group can be trained on appropriate reports.

Sample After Action Reports

After Action Reports (AARs) may be written following any severe weather event where local SKYWARN members are activated. AARs may also be written following exercises. The length and amount of detail of the AAR will depend a lot on the type of event or exercise being covered. A large scale, multi-jurisdiction event such as Hurricane Katrina or the 9/11 attack may require an AAR that is highly detailed, very lengthy, and requires coordinated effort from representatives of all the groups involved. A small scale, localized event may require far less. In either case certain basic information should be included.

Let's take a look at two sample AARs. The first (below), is written for a localized event that required a relatively small response. The second, in the Appendix at the back of the book, is for a communications exercise that involved several groups. You will find that they both contain similar basic information.

To help facilitate a good AAR each participant should be encouraged to write up a brief account of their experience during the event or exercise. The reports can help the writer of the AAR get a better overall picture of the response. One way to do this is to make a simple AAR form, whether paper or online, and ask participants to take a few minutes to fill it out as soon as possible after the event / exercise. For an example of the online AAR report, refer to the Lansing Area / Ingham County Amateur Radio Public Service Corps Web site.[5]

Lafayette County

Amateur Radio Emergency Service

After Action Report

Date of activity: May 3, 2002

Description of activity: SKYWARN activation for severe weather

Duration of activity: 1500-2200 hours

Amateur radio groups participating: Lafayette County ARES / SKYWARN

Served agencies participating: Lafayette County Emergency Management, National Weather Service

Describe served agency participation: Emergency management assisted in gather severe weather reports. All reports of severe weather were forwarded to the National Weather Service. National Weather Service issued weather watches, warnings, and statements.

Number of Amateurs participating: 8

List of amateurs participating: K5DSG, KE5NQP, K5LMB, W5BJC, KD5JHE, N5RB, KE5TMY, WB5VYH

Person-hours of Amateur service: 37

Describe goals of activity, both for served agency and serving group: Local SKYWARN members gathered severe weather reports and submitted them to local net control. Net control then relayed these reports in a timely manner to the National Weather Service via the Amateur Radio station at the National Weather Service Office. Requests to gather specific information were sent from the National Weather Service office via Amateur Radio nets to SKYWARN storm spotters in the local area.

Did the event fulfill the goals? Yes.

What went well? Call up procedures went well, the severe weather net was run effectively.

Areas needing improvement: Additional SKYWARN training is needed, one operator not involved with SKYWARN activities interrupted the net on several occasions, repeater coverage in some areas is not optimal.

Lessons learned: Study repeater coverage area and look at ways to improve, add a relay station to assist net control during times where traffic volume is high, coordinate efforts with neighboring county ARES / SKYWARN, look into future use of *Echolink*.

General comments: Local SKYWARN members did a great job with severe weather reports.

Ideas for future exercises: A severe weather exercise coordinated with local emergency management, district ARES resources, and the National Weather Service would be useful in the near future.

END OF REPORT

What Do We Do With the AAR Once it is Written?

After the AAR has been properly distributed to those that need to read it the AAR should be kept on file. Remember this is a tool for future planning: it must be accessible for future use. One way we can keep an AAR valuable is by conducting a future review. Let's say the AAR was written by the local Emergency Coordinator for a tornado event. Naturally all SKYWARN and ARES members will read it as soon as it is available. And it is likely it may be looked at by the local emergency management director. If it goes on the shelf and is forgotten about at this point then it hasn't served its true purpose. Now let's say a couple months later the SKYWARN and ARES groups are going to conduct an exercise to test their response to severe weather. The EC re-reads the AAR and applies it to how the test is going to be conducted. This helps give some guidance to developing an exercise that encourages learning from real life experience. Now let's say the same area is facing the threat of a tornado a year later. The EC and emergency manager can look over the AAR and find reminders of what worked, what didn't work and eventually see if additional training and exercises paid off.

The AAR is a concept well rooted in emergency management, military and public safety, but it is not necessarily exclusive to these field. Amateur Radio can make valuable use if AARs as a tool for severe weather response.

Debriefing

A debriefing is another useful tool for assessing our response to severe weather. The debriefing typically involves more than just the Amateur Radio storm spotter group. It will likely involve local emergency management and/or public safety officials. It is also possible a debriefing may be done with representatives from the local NWS office.

The purpose of the debriefing is to review the effectiveness of the response and address issues of concern. Being present at the debriefing allows you to directly answer questions that may come up about the Amateur Radio response. It also gives you a forum to bring up issues of concerns for others involved and to ask questions. You will want to bring with you to the debriefing information from the event. Do not rely on memory and log books alone. Throughout the event keep a separate diary of issues for the debriefing session, what *The ARRL Emergency Communications Handbook* refers to as a 'Debriefing Diary'. This should contain issues not appropriate for the station log book, information you will need to retain if the log book has to be handed over to someone else and information about specific events, times, places, etc that need to be mentioned. Here are some other items for the Debriefing Diary:

- What was accomplished?
- Is anything else still pending? Note unfinished items for follow up.
- What worked well? Keep track of things that worked in your favor.
- What needed improvement?
- Ideas to solve known problems in the future.
- Key events.
- Conflicts and resolutions.

Remember to focus on constructive criticism during the debriefing and not attacks on actions taken, finger pointing or casting personal blame.

REFERENCES

[1] Federal Emergency Management Agency, "Emergency Planning", Independent Study IS-235, February 2006.
[2] National Climatic Data Center (NCDC): **www.ncdc.noaa.gov**
[3] ARRL Amateur Radio Public Service page: **www.arrl.org/FandES/field/pubservice.html**
[4] United States Army, "A Leader's Guide to After Action Reviews," September 1993.
[5] Lansing Area / Ingham County Amateur Radio Public Service Corps Web site: **www.lansingarpsc.com**

Appendix 1

Weather Books for the Storm Spotter

General Meteorology / Weather

Weather by Storm Dunlap
Weather: Air Masses, Clouds, Rainfall, Storms, Weather Maps, Climate by Paul Lehr and Will Burnett
The National Audubon Society Field Guide to North American Weather by David Ludlum
The Book of Clouds by John Day
The Weather Wizard's Cloud Book: A Unique Way to Predict the Weather Accurately and Easily by Reading the Clouds by Louis Rubin, J. Duncan and Hiram Herbert
Peterson First Guide to Clouds and Weather by V. Schaefer, P. Pasachoff, R. Peterson
Glossary of Weather and Climate by Ira Greer
Weather Basics by Joseph Basalma and Peter R. Chaston
The AMS Weather Book the Ultimate Guide to America's Weather by Jack Williams

Weather Analysis & Forecasting

Weather Map Handbook by Tim Vasquez
Weather Analysis by Dusan Djuric
Storm Chasing Handbook by Tim Vasquez
Weather Forecasting Handbook by Tim Vasquez
Weather Forecasting: Rules, Techniques and Procedures by George Elliot
Basic Essentials Weather Forecasting by Michael Hodgson
Weather Maps, Third Edition How to Read and Interpret All the Basic Charts by Peter R. Chaston

Specific Storms

Hurricanes! by Peter R. Chaston
Thunderstorms, Tornadoes, and Hail by Peter R. Chaston
Storm Warning: The Story of a Killer Tornado by Nancy Mathis
The Tri-State Tornado: The Story of America's Greatest Tornado Disaster by Peter Felknor
Roar of the Heavens by Stefan Bechtel
Category 5: The 1935 Labor Day Hurricane by Thomas Neil

TEXTBOOKS

Basic

Meteorology Today: An Introduction to Weather, Climate, and Environment by C.D. Ahrens
A World of Weather Fundamentals of Meteorology by Jon Nese and Lee Grenci

Advanced

Meteorology for Scientists and Engineers by Roland Stull
Mid-Level Weather Systems by Toby Carlson
Mesoscale Meteorology and Forecasting by Peter Ray

Appendix 2

Weather Web Sites

National Weather Service
National Weather Service Main Page — www.weather.gov
Interactive National Weather Service — http://inws.wrh.noaa.gov
E-Spotter — http://espotter.weather.gov
JetStream — www.srh.noaa.gov/jetstream/
NWS Aware — www.nws.noaa.gov/os/Aware/index.shtml
Storm Prediction Center — www.spc.noaa.gov
National Hurricane Center — www.nhc.noaa.gov
Central Pacific Hurricane Center — www.prh.noaa.gov/hnl/cphc
National Severe Storms Laboratory — www.nssl.noaa.gov
SKYWARN — www.nws.noaa.gov/skywarn
NOAA Weather Radio All Hazards — www.nws.noaa.gov/nwr

Commercial Sites
Skype — www.skype.com
Google Earth — http://earth.google.com
AccuWeather.com — www.accuweather.com
WeatherBug — http://weather.weatherbug.com
Storm Pulse — www.stormpulse.com
Weather Underground — www.wunderground.com
The Weather Channel — www.weather.com
Spotter Network — www.spotternetwork.org
Weather Tap — www.weathertap.com
WX Spots — www.wxspots.com
Weather Defender — www.weatherdefender.com
GRLevel X — www.grlevelx.com
Allison House — www.allisonhouse.com
Storm Alert — www.interwarn.com

Organizations
SKYWARN — www.skywarn.org
American Meteorological Society — www.ametsoc.org
National Weather Association — www.nwas.org

Education
University of Illinois WW2010 — http://ww2010.atmos.uiuc.edu/%28Gh%29/home.rxml
University Corporation for Atmospheric Research — www2.ucar.edu

Amateur Radio and CWOP
American Radio Relay League — www.arrl.org
APRS — www.aprs.org
Echolink — www.echolink.org
D-Star — www.dstarusers.org
The Hurricane Watch Net — www.hwn.org
Citizens Weather Observation Program — www.wxqa.com

Storm Spotting & Amateur Radio

INDEX TO SKYWARN WEB PAGES ON THE INTERNET

Compiled by Todd L Sherman, KB4MHH, Coordinator, Alachua County SKYWARN

General SKYWARN Sites
National SKYWARN Home Page　　　　　　www.skywarn.org
(NOTE: This is a personal Web page, not officially funded, sponsored, or run by the USDoC, NOAA, or the NWS.)
REACT International　　　　　　　　　　www.reactintl.org

SKYWARN Sites by State

Alabama:
Northern Alabama SKYWARN　　　　　　http://nalsw.net
NWS Birmingham SKYWARN　　　　　　www.srh.noaa.gov/bmx/skywarn
Walker Co SKYWARN　　　　　　　　　www.qsl.net/n4emp/about.html

Alaska:
None yet found

Arizona:
Maricopa County SKYWARN　　　　　　http://azskywarn.weatherbus.com
NWS Flagstaff SKYWARN Page　　　　　www.wrh.noaa.gov/fgz/information/spotter/spotter_page.php?wfo=fgz

Arkansas:
Arkansas SKYWARN　　　　　　　　　www.carenclub.com/skywarn.php
Diamond Lakes Area SKYWARN　　　　www.qsl.net/w5lvb/skywarn
Northeast Arkansas SKYWARN Network　www.neaskywarn.org
NE Oklahoma/NW Arkansas SKYWARN　www.okarkskywarn.org

California:
Southwest California SKYWARN　　　　http://swskywarn.org
San Francisco Bay Area SKYWARN　　　www.wrh.noaa.gov/mtr/spotter.php

Colorado:
Tri-State SKYWARN (KS / NE / CO)　　　www.crh.noaa.gov/gld/?n=/skywarn/index.php

Connecticut:
Connecticut SKYWARN　　　　　　　　www.ctskywarn.co
New London County SKYWARN　　　　www.99main.com/~kd1ld/skywarn.html

Deleware:
Delaware SKYWARN /
Mt Holly NWSFO / WX4PHI　　　　　　http://skywarn.wordpress.com

District of Columbia:
NWS Washington / Baltimore SKYWARN　www.nws.noaa.gov/er/lwx/skywarn/skywarn.htm

Florida:
Alachua County SKYWARN Home Page　http://alachuaskywarn.org
Calhoun County SKYWARN　　　　　　http://community.webtv.net/jimstormspoter/SKYWARNOFNWFLA
Collier County SKYWARN　　　　　　　www.colliergov.net/Index.aspx?page=1102
Flagler County SKYWARN　　　　　　　www.react4800.org/skywarn.htm
Glade SKYWARN　　　　　　　　　　　www.gladeskywarn.org
Jacksonville SKYWARN Association　　www.nofars.org/skywarn.html
Marion County ARES　　　　　　　　　www.qsl.net/mcares/index.html
Martin County SKYWARN　　　　　　　www.gate.net/~ke4uei

Storm Spotting & Amateur Radio

Northeast Florida SKYWARN	http://groups.yahoo.com/group/NEFLSkywarn
NWS-JAX SKYWARN Page	www.srh.noaa.gov/jax/skywarn.shtml
Osceola Co ARES/SKYWARN	www.osceolacountyares.org/?skywarn
Palm Beach Co SKYWARN	www.pbcskywarn.org
Pinellas County SKYWARN	www.pcacs.org/skywarn.htm
Seminole County SKYWARN	www.seminoleares.org/skywarn.html
St Johns County ARES / SKYWARN	www.saarsham.net/areshome.html
St Lucie County SKYWARN	www.qsl.net/slcskywarn
NWS-Tallahassee SKYWARN	www.srh.noaa.gov/tae/severe.php
Treasure Coast Weather Network	www.tcwn.net

Georgia:

Albany Amateur Radio Club	www.w4mm.com
Clinch County SKYWARN	http://twister31634.tripod.com
Georgia ARES / SKYWARN Home Page	www.georgiaskywarn.com
Gwinnett County SKYWARN	www.gwinnettskywarn.com
N4UYB's Georgia SKYWARN	www.geocities.com/becky63.geo/GaSkyGA.html
Hall County SKYWARN Todd Bowden, AF4CZ	www.lanierlandarc.org/hall_skywarn.htm
Northeast Georgia SKYWARN Ed Gilbert	http://negarcskywarn.netforms.com

Hawaii:

WFO Honolulu SKYWARN	www.prh.noaa.gov/hnl/skywarn

Idaho:

Eastern Idaho SKYWARN Spotter Program	www.wrh.noaa.gov/pih/Spotter/spottersched.php
NWS Boise SKYWARN Spotter Program	www.wrh.noaa.gov/boi/awareness/skywarn.php

Illinois:

Bolingbrook SKYWARN	www.k9bar.org
Bureau County SKYWARN	www.bureaucountyskywarn.net
Central Illinois SKYWARN	www.w9nws.com
DuPage Amateur Radio Club SKYWARN	http://skywarn.w9dup.org
Illinois area SKYWARN Spotters Blog	www.ilskywarn.com
Jefferson/Wayne County SKYWARN	http://arcomarc.tripod.com/skywarn.htm
Lake County ARES / RACES / SKYWARN	www.lakecountyskywarn.org
Multi-County SKYWARN of Northern Illinois	http://weather.cod.edu/skywarn
Winnebago County ARES / SKYWARN	www.winnebagoares.org

Indiana:

Dayton Section SKYWARN   www.dayton-skywarn.org (in Indiana covering the counties of Auglaize, Champaign, Hardin, Logan, Mercer, Shelby. Also covering some counties in Ohio (see).

Elkhart County SKYWARN	www.imoskywarn.org
Knox County SKYWARN	http://w9eoc.tripod.com/swpg1.htm
Pike County SKYWARN	http://members.tripod.com/~pcskywarn
South Central Indiana SKYWARN	www.geocities.com/kb9jlf
Wayne County SKYWARN	www.joesproject.com/skywarn

Iowa:

Iowa-Skywarn.org	www.iowa-skywarn.org
Iowa Storm Spotters	www.raccoon.com/~pcraven/storm/ia_spotters.html
NWSFO Des Moines SKYWARN Program	www.crh.noaa.gov/dmx/?n=skywarn

Kansas:

Johnson Co Emergency Comms Service, Inc	www.k0ecs.com
Wichita NWS-SKYWARN	www.ict-skywarn.org

Storm Spotting & Amateur Radio

Kentucky:
Southeast Kentucky SKYWARN — www.freewebs.com/sekyskywarn
MesoTrack SKYWARN — www.mesotrack.com

Louisiana:
Lafayette SKYWARN — http://info.louisiana.edu/mahler/skywarn
Southeast Louisiana SKYWARN — www.selaskywarn.org

Maine:
WX1GYX SKYWARN — www.wx1gyx.org

Maryland:
NWS Washington / Baltimore SKYWARN — www.nws.noaa.gov/er/lwx/skywarn/skywarn.htm

Massachusetts:
Cape Cod and the Islands SKYWARN Page — http://radiotimeline.com/n2knl.htm
Eastern Massachusetts
ARES RACES and SKYWARN Homepage — www.ultranet.com/~rmacedo

Michigan:
Branch County SKYWARN — www.bronsonweather.org/skywarn
Lansing Area SKYWARN — www.lansingarpsc.com/skywarn.html
Macomb County Amateur Radio Public Service
Corp (ARES / RACES / SKYWARN) — www.wa8mac.net
Oakland County ARPSC
(SKYWARN / RACES / ARES) — www.arpsc.com/skywarn.html
Newaygo County, West Michigan — www.tdats.net
NWS-Gaylord Spotter Page — www.crh.noaa.gov/apx/spotters.php
NWS White Lake Spotter's Page — www.crh.noaa.gov/dtx/skywarn.php
Washtenaw County SKYWARN — www.ewashtenaw.org/government/departments/emergency_management/em_severe.html
West Michigan SKYWARN — www.kb8sew.net/cgi-bin/skywarn.pl

Minnesota:
Duluth, MN SKYWARN Program — www.crh.noaa.gov/dlh/prepare/skywarn/skywarn.php
Isanti County SKYWARN — www.isantiskywarn.org
Minn / St Paul Metro Area SKYWARN Central — www.metroskywarn.org
NWS Twin Cities / Chanhassen — www.crh.noaa.gov/mpx/skywarn.php

Mississippi:
North Central Mississippi SKYWARN — www2.msstate.edu/~bmm2/gplains.htm
South Mississippi Emergency Repeater Group — http://smerg.org

Missouri:
Missouri SKYWARN Repeater List — www.qsl.net/kb0rpj/skywarn
Mid-Missouri SKYWARN Assn — www.midmoskywarn.net
Pulaski County ARES / SKYWARN — www.angelfire.com/mo3/emergency/ARES1.html
Mokan SKYWARN of SW Missouri — www.mokanep.com
St Louis County SKYWARN / RACES — www.stlouisskywarn.com

Montana:
Big Sky SKYWARN — www.w7eca.org/SKYWARN

Nebraska:
Heartland REACT — http://heartlandreact.freehosting.net
Southwest Nebraska (& NW Kansas) SKYWARN — http://stormchaser231.tripod.com
Tri-State (KS, NE, CO) SKYWARN Goodland, KS — www.crh.noaa.gov/gld/?n=/skywarn/index.php

Storm Spotting & Amateur Radio

Nevada:
Las Vegas Regional SKYWARN   www.qsl.net/lvr-skywarn
SKYWARN Las Vegas   www.wrh.noaa.gov/vef/skywarn.php (covering Mohave Co Arizona,
    Sanbernardino and Inyo Counties, California, and Clark, Esmeralda, Nye and Lincoln Counties, Nevada.)

New Hampshire:
WX1GYX SKYWARN   www.wx1gyx.org

New Jersey:
Bergen County SKYWARN   www.bergenskywarn.org
Cumberland County SKYWARN   www.cumberlandskywarn.com
MercerCounty SKYWARN   www.freewebs.com/merskywarn
Lucas County SKYWARN   www.LucasCountySkywarn.org
Warren County SKYWARN   www.wc2em.net/Warren%20County%20Skywarn.htm

New Mexico:
Chaves County SKYWARN   http://roswelltornado.com
Eddy County SKYWARN   http://eddyoem.com/skywarn.html
NWS Albuquerque SKYWARN   www.srh.noaa.gov/abq/skywarn/skywarn.htm
Southeast New Mexico SKYWARN   http://kd5zmd88232.tripod.com

New York:
NWSFO Albany SKYWARN Home Page   www.nws.noaa.gov/er/okx/Skywarn/skywarn.html
NWS-Buffalo SKYWARN Page   www.wbuf.noaa.gov/spt.htm
Oswego County RACES / SKYWARN   www.oswegoraces.org
Town of Babylon SKYWARN   www.tobares.org/skywarn.html
Westchester County Skywarn   www.weca.org/skywarn.html

North Carolina:
Central Carolina SKYWARN   www.centralcarolinaskywarn.net
Newport District SKYWARN   www.mhxskywarn.org
FCARCS SKYWARN Home Page   www.geocities.com/TheTropics/3212/skywarn.html
North Carolina SKYWARN   www.ncarrl.org/ares/skywarn
North Carolina Triad SKYWARN   www.triadskywarn.com
NWS-Raleigh SKYWARN Page   www.erh.noaa.gov/rah/skywarn

North Dakota:
Mountrail County SKYWARN Chapter Homepagewww.mountrailskywarncoc.cjb.net
SKYWARN North Dakota   www.crh.noaa.gov/bis/?n=skywarn
North Dakota SKYWARN   http://groups.yahoo.com/group/WardCountyRACES

Ohio:
Sandusky / Ottawa County SKYWARN   www.aressc.org
Central Ohio Severe Weather Network www.severe-weather.org
Cleveland SKYWARN   www.erh.noaa.gov/cle/skywarn/skywarn.html
Cuyahoga County SKYWARN   www.ccsw.us
Dayton Section SKYWARN   www.dayton-skywarn.org (in Ohio, covering the counties of Clark,
    Darke, Fayette, Greene, Miami, Montgomery, Preble, Union, Northern Warren, and Wayne. Also covering some counties
    in Indiana (see.)
Lorain County SKYWARN   www.loraincountyskywarn.org
Lucas County SKYWARN   www.LucasCountySkywarn.org
North Central Ohio SKYWARN   www.ncos.org
Northwest Ohio SKYWARN   www.nwohioskywarn.org
Summit County SKYWARN   www.sumsky.us
Weather Amateur Radio Network (WARN)
(Cincinnati, OH SKYWARN)   www.warn.org

Appendix 2

Storm Spotting & Amateur Radio

Oklahoma:
Altus SKYWARN Association	www.altusskywarn.org
Carter County SKYWARN	www.cartercountyskywarn.org
Blackwell SKYWARN	www.blackwellskywarn.org
Cleveland County SKYWARN	www.qsl.net/clevelandcoskywarn
NWSFO Norman SKYWARN and Spotter Page	www.srh.noaa.gov/oun/skywarn
Northeast Oklahoma / Northwest Arkansas SKYWARN	www.okarkskywarn.org
Southern Oklahoma SKYWARN Group	www.soaresss.org

Oregon:
Medford Area NWS SKYWARN	www.emcomm.org/skywarn
Portland SKYWARN	www.wrh.noaa.gov/pqr/skywarn.php

Pennsylvania:
Delaware County SKYWARN	www.delcoskywarn.net
Erie County ARES / SKYWARN	http://w3gv.org/erie-county-skywarn
Johnstown SKYWARN	http://johnstownpaweather.com/skywarn.html
NWS Pittsburgh, PA SKYWARN	www.nws.noaa.gov/er/pit/skywarn.htm
Schuylkill County SKYWARN	www.w3sc.org/#S-warn

Rhode Island:
Northern Rhode Island REACT / SKYWARN	www.ultranet.com/~nrireact
Rhode Island SKYWARN	http://riskywarnteam.8m.net

South Carolina:
Horry County SKYWARN (Myrtle Beach)	www.w4gs.org/ARES/skywarn.htm
Laurens County SKYWARN	www.backroads.net/~k4kep/SKYWARN
NWS Greenville SKYWARN	www.nws.noaa.gov/er/gsp/skywarn/skywarn.htm
Pickens County SKYWARN	www.qsl.net/kq4ao/skywarn.html

South Dakota:
NWS Rapid City SKYWARN	www.crh.noaa.gov/unr/?n=skywarn

Tennessee:
East Tennessee SKYWARN (District 5 and 6)	www.tnskywarn.com
Lawrence Co SKYWARN	www.lawrencecountyskywarn.org
NETN District 7 SKYWARN	www.wx4tn.org

Texas:
Dallas County / City RACES / SKYWARN	www.dallasraces.org
Deep East Texas SKYWARN	www.stormwarn.net
East Texas SKYWARN	http://skywarn.n4ki.us
East Texas Storm Spotters	www.lamardenby.orgetss.htm
South Plains Storm Spotting Team, Inc	www.spsst.org
Midland, Texas SKYWARN	www.srh.noaa.gov/maf/?n=top_skywarn
Parker County SKYWARN	www.homestead.com/K5DC/wx.html
Potter-Randall County ARES / RACES	www.qsl.net/nwtx-ares
Texas SKYWARN Home Page	www.skywarn-texas.org
Tulsa Amateur Radio Club Weather Spotters	http://w5ias.com

Utah:
Nothing yet found

Vermont:
NWS Burlington VT SKYWARN Home Page	www.erh.noaa.gov/btv/html/skywarn/skywarn.shtml

Virginia:
Blacksburg, VA SKYWARN — www.wx4rnk.org
Sterling Virginia SKYWARN Program — www.147300.com

Washington:
NWS Seattle SKYWARN — www.wrh.noaa.gov/sew/spotter.php

West Virginia:
Charleston SKYWARN — www.nws.noaa.gov/er/rlx/skywarn/skywarn.html
Mingo Co ARES / RACES / SKYWARN — www.qsl.net/mingoares/index.html

Wisconsin:
Milwaukee Area SKYWARN Assn — www.mke-skywarn.org
Outagamie County SKYWARN — http://skywarn.webhop.net
Waupaca Co ARES / RACES / SKYWARN — www.waupacacountyares.org

Wyoming:
Southeast Wyoming SKYWARN — www.wyomingskywarn.net

SKYWARN SITES ELSEWHERE

CANADA:
CanWarn Home Page — http://on.ec.gc.ca/canwarn/home-e.html
Edmonton CANWARN (Alberta) — www.angelfire.com/philccom/canwarn.html
Frontenac County Amateur Radio Emergency Services Group CANWARN (E Ontario) — http://fcares.net/canwarn.html
Mississauga CANWARN (Ontario) — www.marc.on.ca/marc/services/serv_canwarn.asp
Kitchener-Waterloo CANWARN (Ontario) — www.kwarc.org/canwarn/index.htm
CANWARN Hamilton (Ontario) — www.canwarnhamilton.ca

PUERTO RICO:
Puerto Rico SKYWARN — www.skywarnpr.org

(This list created June 8, 1996, last updated November 11, 2009. Copyright © 1996 - 2009 Todd L Sherman. All Rights Reserved. Published with permission.)

Appendix 3

A Local SKYWARN Operations Manual

At the local level a SKYWARN group may develop a local operations manual. Doing so will add some structure and strength to the organization. It also helps local SKYWARN members to know what to do, who to report to, and how local net operations are conducted. Alachua County (Florida) SKYWARN developed such a manual in June 2000. While some simple basics were borrowed from a very early version of the *ARRL Emergency Coordinator's Handbook* (1986 edition) and elsewhere, the manual was designed largely using NASA's Mission Control Center in Houston, Texas as a model. This idea allows for tasks to be relegated or assigned and permits multiple things to be done at the same time simply by giving specific jobs to specific positions. Basically, many of the tasks that an NCS might be required to take on could be assigned to others so they can be accomplished faster and much more efficiently. Relegating tasks to others relieves pressure on the NCS and helps to alleviate chances for confusion, and frustration, and lowered morale from taking on a task that can get very tedious when things start to get fast-paced, and increases speed of situation handling.

Developing a local operations manual can be one of the biggest challenges a local SKYWARN group can face. Severe weather isn't the same everywhere and neither is the response. For example, some areas may have civilian organizations and others may have started within local Emergency Management. Others may even be a part of a military structure – say, operating under a MARS-related organization. (Some MARS organizations have also used the Alachua County SKYWARN SOP manual as a template.) For this reason, it is impossible to simply have one operations manual that will work for all locations. The manual developed in Alachua County, however, does give us a template to work with in developing a local manual.

The manual is divided into two sections: content and appendices. Content covers the basic information about the organization, membership, training, activation, net control, and safety. The appendices are a reference section for storm spotters. Let's go over what a model operations manual will cover.

CONTENT

Purpose – This is a statement that addresses the purpose of SKYWARN and the local SKYWARN group. It also covers the purpose of the manual.

Limitations – This sets out limits within the organization. It may spell out limits placed on the manual, members, and purposes of the organization. This is also where we would clearly state limitations on mobile spotters to not act as storm chasers.

Organizational Structure – Local SKYWARN organizational structure should be clearly spelled out. Positions included may include: SKYWARN Coordinator, Assistant SKYWARN Coordinator, training managers, net managers, public information, recruiting, and technical managers. The number of positions will depend upon the size of the SKYWARN group.

Membership – Indicate membership requirements such as Amateur Radio license, NWS SKYWARN training, membership cards, etc.

Training – Required and expected training should be covered here. Don't forget to look beyond the SKYWARN class, cover training nets, local training meetings, SETs, and any other training functions.

Alert Authorization – This should clearly indicate how the local SKYWARN group is activated and who has the authority to activate the group. Remember that there are times that severe weather will develop with no little to no warning. Be sure to address activation procedures from local emergency management or SKYWARN coordinator. Also include information on the repeater to be used for SKYWARN activation.

Mobilization – This covers how members will be advised of SKYWARN activation. Include all methods that may be used: radio, telephone, cell phone, pager, email, etc. Also address what SKYWARN members are to do when mobilized, typically this will be to check into the SKYWARN net and await further instructions. Another topic within mobilization will be levels of activation such as stand by alert or emergency alert. Your group may use other indicators such as conditions green, yellow, and red.

Operation Procedures – This section explains how operations are to be conducted during SKYWARN activation. Include specific operation instructions for each level of alert, reporting procedures, how and when the net will be identified, and how and when the net will be secured.

Net Procedures – The SKYWARN net can operate like a well oiled machine or can be total chaos. How you net goes depends on how well it is structured and the instructions provided for net control stations and stations checking in. A portion of your manual should include all instructions for conducting a directed severe weather net. There are two areas that should be addressed. First address net control stations. Second address the storm spotters that will be checking in.

Safety – Safety must be taken seriously at all times. The local SKYWARN group should have a designated Spotter Safety Officer. The manual should spell out the SSO's duties which typically include: monitor radar activity, monitor spotter reports to identify safety issues in the area, plot these events on a map, and when necessary advise net control of dangerous situations for spotters and provide information on the safest and fastest route away from the danger area. Additionally the SSO will monitor for unsafe activity from spotters. This information should be provided to the SKYWARN Coordinator or other designated individual for follow up action.

Frequencies – The manual should list the specific severe weather net frequencies that will be used by the local SKYWARN group. List only frequencies used for local SKYWARN communication; primary repeater or simplex frequency and any back up frequency should be listed. Frequencies for NOAA Weather Radio, area repeaters, public safety, etc. should be listed in an appendix.

Definitions and Miscellaneous – List here any acronyms or terms that are frequently used by the group and their meaning. Also provide definitions for NWS terms e.g. watches, warnings, and advisories. Reportable criteria may also be listed here. This may also be a place where you can list "don't do" items.

APPENDICES

The appendices is a reference section for the SKYWARN manual. Specific contents of the appendices will vary but here are a few ideas that you may want to include.

Related Government Contacts – Include local, state, and NWS contacts. Also include served agencies such as the Red Cross and Salvation Army.

Storm Spotter Roster – Include names, call signs, and contact information.

Net Preambles – A script for opening the net, net continuity, announcements, roll call, and securing the net.

Phonetic Alphabet

Frequencies – local SKYWARN frequencies, NOAA Weather Radio frequencies, any simplex frequencies used.

Spotter's Checklist – Items that a spotter may need, especially for mobile spotting.

Spotter Locations – List known good spotting locations. Be sure to emphasize spotter safety.

Cities in the Area – List area towns and cities and their respective counties.

Maps – Provide any appropriate maps: city, county, district, etc. Don't forget to include your local NWS County Warning Area.

Reporting Form – A form listing specific reportable weather events.

Net Control Log Form – This log sheet is used by net control to keep track of net activity. It should include: Called in time, call sign, name, time in, time out, location, and comments.

Charts – Beaufort Scale, Fujita Scale, Saffir-Simpson Scale, hail size chart.

Recommended Books – Books for the storm spotter

SKYWARN Products – caps, jackets, badges, magnetic signs, etc.

Revision History – Make sure to indicate any revisions to the manual; what was revised, when, and any other pertinent information.

Appendix 4

Memorandum of Understanding
between the National Weather Service and the American Radio Relay League, Inc.

I. PURPOSE

The purpose of this document is to state the terms of a mutual agreement (Memorandum of Understanding) between the National Weather Service (NWS) and the American Radio Relay League, Inc. (ARRL), that will serve as a framework within which volunteers of the ARRL may coordinate their services, facilities, and equipment with NWS in support of nationwide, state, and local early weather warning and emergency communications functions. It is intended, through joint coordination and exercise of the resources of ARRL, NWS, and Federal, State and local governments, to enhance the nationwide posture of early weather warning and readiness for any conceivable weather emergency.

II. RECOGNITION

The National Weather Service recognizes that the ARRL is the principal organization representing the interests of more than 400,000 U.S. radio-amateurs and because of its Field Organization of trained and experienced communications experts, can be of valuable assistance in early severe weather warning and tornado spotting.

The American Radio Relay League recognizes the National Weather Service with its statutory responsibility for providing civil meteorological services for the people of the United States. These services consist of:

1. Issuing warnings and forecasts of weather and flood conditions affecting the nation's safety, welfare and economy; and,
2. Observing and reporting the weather of the U.S. and its possessions.

To perform these functions and many related, specialized weather services, NWS operates a vast network of stations of many types within the U.S.; it cooperates in the exchange of data in real time with other nations, including obtaining of weather reports from ships at sea.

I. PURPOSE

The purpose of this document is to state the terms of a mutual agreement (Memorandum of Understanding) between the National Weather Service (NWS) and the American Radio Relay League, Inc. (ARRL), that will serve as a framework within which volunteers of the ARRL may coordinate their services, facilities, and equipment with NWS in support of nationwide, state, and local early weather warning and emergency communications functions. It is intended, through joint coordination and exercise of the resources of ARRL, NWS, and Federal, State and local governments, to enhance the nationwide posture of early weather warning and readiness for any conceivable weather emergency.

II. RECOGNITION

The National Weather Service recognizes that the ARRL is the principal organization representing the interests of more than 400,000 U.S. radio-amateurs and because of its Field Organization of trained and experienced communications experts, can be of valuable assistance in early severe weather warning and tornado spotting.

The American Radio Relay League recognizes the National Weather Service with its statutory responsibility for providing civil meteorological services for the people of the United States. These services consist of:

1. Issuing warnings and forecasts of weather and flood conditions affecting the nation's safety, welfare and economy; and,
2. Observing and reporting the weather of the U.S. and its possessions.

To perform these functions and many related, specialized weather services, NWS operates a vast network of stations of many types within the U.S.; it cooperates in the exchange of data in

expert assistants to administer the various emergency communications and public service programs in the section. Each section has a vast cadre of volunteer appointees to perform the work of Amateur Radio at the local level, under the supervision of the Section Manager and his/her assistants.

IV. ORGANIZATION OF THE NATIONAL WEATHER SERVICE

The National Weather Service consists of a National Headquarters in Washington, D.C., and six regional offices in the United States: Eastern, Southern, Central, Western, Alaska, and Pacific. An NWS Public Information Office is located at Weather Service Headquarters. Fifty-two Weather Service Forecast Offices and 209 Weather Service Offices provide warnings and forecasts to the Nation.

SKYWARN is the spotter program sponsored by the NWS. Radio amateurs have assisted as communicators and spotters since its inception. In areas where tornadoes and other severe weather have been know to threaten, NWS recruits volunteers, trains them in proper weather spotting procedures and accepts the volunteers' reports during watches and episodes of severe weather. By utilizing the SKYWARN volunteers, the NWS has "eyes and ears" throughout the affected area in conjunction with NWS sophisticated weather monitoring equipment.

V. PRINCIPLES OF COOPERATION

A. The American Radio Relay League agrees to encourage its volunteer Field Organization appointees, especially the Amateur Radio Emergency Service, to contact and cooperate with Regional Weather Service Headquarters for the purpose of establishing organized SKYWARN networks with radio amateurs serving as communicators and spotters.

B. ARRL further agrees to encourage its Section management teams to provide specialized communications and observation support on an as-needed basis for NWS offices in other weather emergencies such as hurricanes, snow and heavy rain storms, and other severe weather situations.

C. The National Weather Service agrees to work with ARRL Section Amateur Radio Emergency Service volunteers to establish SKYWARN networks, and/or other specialized weather emergency alert and relief systems. the principle point of contact between the ARRL Section and local NWS offices is the Meteorological Services Division of the appropriate NWS Regional Office. The addresses of the Regional offices are listed below. The national contact for ARRL is the Public Service Branch, ARRL Headquarters, Newington, CT 06111.

National Weather Service Eastern Region
NOAA
585 Stewart Avenue
Garden City, New York 11530
Tel: 516-228-5400

National Weather Service Southern Region
NOAA
819 Taylor Street, Rm. 10A26
Fort Worth, Texas 76102
Tel: 817-334-2668

National Weather Service Central Region
NOAA
601 E. 12th St., Rm. 1836
Kansas City, Missouri 64106
Tel: 816-374-5463

National Weather Service Western Region
NOAA
Box 11188, Federal Bldg.
125 S. State Street
Salt Lake City, Utah 84147
Tel: 801-524-5122

National Weather Service Alaska Region
NOAA
Box 23, 701 C St.
Anchorage, Alaska 99513
Tel: 907-271-5136

National Weather Service Pacific Region
NOAA
P.O. Box 50027
Honolulu, Hawaii 96850
Tel: 808-546-5680

Page last modified: 08:16 AM, 20 Mar 2001 ET
Page author: rwhite@arrl.org
Copyright © 2001, American Radio Relay League, Inc. All Rights Reserved

Appendix 5

ARRL Reporting Forms

PUBLIC SERVICE ACTIVITY REPORT

About This Form

Amateur Radio donates thousands of person hours of supplementary public service communications in civil emergencies, official drills and events such as parades and marathons each year. Such events show Amateur Radio in its best light, and it is critically important that ARRL bring documentation of this public service work to the attention of the Congress, the FCC and other public officials. Your information below is an important addition to the record. Please complete and return this form to the Public Service Branch at ARRL Headquarters. Thank you.

1. Nature of activity (Check one).

__ *Communications Emergency*. Amateurs supplied communications required to replace or supplement normal communications means.

__ *Alert*. Amateurs were deployed for emergency communications, but emergency situation did not develop.

__ *Special exercise*. Amateurs supplied communications for a parade, race, etc.

__ *Test or drill*. A training activity in which amateurs participated.

2. Brief description of activity: _____

3. Places or areas involved: _____

4. Number of amateurs participating: _____

5. Event start date/time: _____ 6. Event end date/time: _____

7. Duration of event (hours): _____ 8. Total person-hours: _____

9. Number of repeaters used: _____

Storm Spotting & Amateur Radio

10. Estimated person-power cost: $ _____ (person-hours times $19/hr)

11. Estimated cost of equipment used: $ _____ (hand-helds, repeaters, etc.)

12. Total estimated cost of service: $ _____ (add amounts from lines 10 and 11)

13. Nets and/or frequencies used (including repeater call signs):
_____
_____

14. Number of messages handled: _____

15. Names of agencies receiving communications support:
_____
_____
_____
_____

16. Please list call signs of amateurs who were major participants:
_____
_____
_____
_____

17. Other comments:
_____
_____
_____
_____
_____
_____

Please attach photos of amateurs in action, newspaper clippings or other data.

Name of Amateur Radio organization providing service:
_____

Location of organization: City or town: _____ State: _____

Your name: _____ Call sign: _____

Address: _____

ARRL appointment, if any: _____ e-mail address: _____

Storm Spotting & Amateur Radio

Monthly DEC/EC Report

Jurisdiction: _____ Month: _____ Year _____

AMATEUR RADIO EMERGENCY SERVICE

Total number of ARES members: _____   Change since last month: _____ (+,-, or same)

Local Net Name: _____ Total sessions _____

NTS liaison is maintained with the _____ Net

Number of drills, tests and training sessions this month: _____   Person hours _____

Number of public service events this month: _____   Person hours _____

Number of emergency operations this month: _____   Person hours _____

Total number of ARES operations this month: _____   Total Person hours _____

Comments:

Signature: _____ Title: (EC or DEC) _____ Call sign: _____
Please send to your SEC or DEC as appropriate by 2nd of the month   FSD-212 (1-04)

Appendix 5

Storm Spotting & Amateur Radio

Monthly Section Emergency Coordinator Report to ARRL Headquarters

ARRL Section: _____ Month: _____ Year _____

AMATEUR RADIO EMERGENCY SERVICE

Total number of ARES members: _____ Change since last month: _____ (+, -, or same)

# of DECs/ECs reporting this month: _____ # of ARES nets active: _____ # with NTS liaison: _____

Calls of DECs/ECs reporting: _____

Number of drills, tests and training sessions this month: _____ Person hours _____

Number of public service events this month: _____ Person hours _____

Number of emergency operations this month: _____ Person hours _____

Total number of ARES operations this month: _____ Total Person hours _____

Comments:

Signature: _____ Call sign: _____

Please send to ARRL HQ, 225 Main St., Newington, CT 06111 by 10th of the month   FSD-96 (1-04)

Storm Spotting & Amateur Radio

2009 EC Annual Report
FORM C

PLEASE PRINT
Based upon your current file of registration cards (FSD-98) for present Amateur Radio Emergency Service members within your jurisdiction, please compile the following information. Information from this form is used to indicate general trends in ARES activity. The more reports returned, the more accurate our analysis will be. Please take a few minutes to complete this report. Mail one copy to ARRL Headquarters, one copy to your SEC/DEC and retain one copy for your own files.

ARRL Section:	
Area of Jurisdiction:	
Name and Call:	
E-mail address:	

#		
1	Total number of ARES members in your group	
2	Total number of new licensees (licensed since 2005) in your ARES group	
3	Total number of members who operate CW	
4	Total number of members who operate VHF	
5	Total number of members with HF emergency-power capability at home	
6	Total number of members with VHF emergency-power capability at home	
7	Total number of members who can operate HF mobile	
8	Total number of members who can operate VHF mobile	
9	Total number of members who are active on packet radio	
10	Approximate number of ARES drills or nets per year	
11	Is your ARES net affiliated with or have liaison to the ARRL National Traffic System?	
12	List the primary agencies served by your ARES group (civil preparedness, Red Cross, NWS, etc)	
13	Does your ARES group work with RACES? If yes, explain the relationship	

Deadline for reporting is February 1, 2010
Mail to: ARRL HQ, 225 Main St, Newington, CT 06111, or e-mail the form to <u>sewald@arrl.org</u>

CC: ARRL HQ, SEC/DEC

Appendix 6

24.00 *FALSE STATEMENTS*

24.01 *STATUTORY LANGUAGE: 18 U.S.C. § 1001*

§1001. *Statements or entries generally*

(a) . . . [W]hoever, in any matter within the jurisdiction of the executive, legislative, or judicial branch of the Government of the United States,[1] knowingly and willfully —

> (1) falsifies, conceals or covers up by any trick, scheme, or device a material fact;
>
> (2) makes any materially false, fictitious, or fraudulent statements or representation; or
>
> (3) makes or uses any false writing or document knowing the same to contain any materially false, fictitious, or fraudulent statement or entry;
>
> shall be fined under this title or imprisoned not more than 5 years . . . .[2]

---

1. The *False Statements Accountability Act of 1996*, Pub. L. No. 104-292, 110 Stat. 3459, changed the language of Section 1001, which previously criminalized false statements made "in any matter within the jurisdiction of any department or agency of the United States . . . [.]" The *False Statements Accountability Act* superseded the Supreme Court's 1995 decision in *Hubbard v. United States*, 514 U.S. 695, 702-03 (1995), which held that the previous version of Section 1001 prohibited only false statements made to the executive branch. The *False Statements Accountability Act* extended the application of Section 1001 to false statements or entries on any matter within the jurisdiction of the executive, legislative or judicial branch of the federal government. However, this prohibition does not apply to a party to a judicial proceeding, or to that party's counsel, "for statements, representations, writings or documents submitted by such party or counsel to a judge or magistrate in that proceeding." 18 U.S.C. § 1001(b).

2. The *Intelligence Reform and Terrorism Prevention Act of 2004*, Pub. L. No. 108-458, 118 Stat. 3638, with an effective date of December 17, 2004, increased the penalties under Section 1001 for crimes involving international or domestic terrorism to include a term of imprisonment of not more than 8 years. Two separate pieces of legislation, each of which would increase the term of imprisonment under Section 1001 for crimes involving terrorism to not more than 10 years, are currently pending in Congress. See *Counter-Terrorism and National Security Act of 2007*, H.R. 3147, 110th Cong. (1st Sess. 2007); *Violent Crime Control Act of 2007*, H.R. 3156, 110th Cong. (1st Sess. 2007).

Under 18 U.S.C. § 3571, the maximum fine under Section 1001 is at least $250,000 for individuals and $500,000 for corporations. Alternatively, if any person derives pecuniary gain from the offense, or if the offense results in a pecuniary loss to a person other than the defendant, the defendant may be fined not more than the greater of twice the gross gain or twice the gross loss. The purpose of Section 1001 is "to protect the authorized functions of governmental departments and agencies from the perversion which might result from" concealment of material facts and from false material representations.

Appendix 7

Integrating
Google Earth, NWS Data and APRS Using KML
by Robert Andrews, K0RDA

I wanted to find an easy way to integrate mapping, radar data, and APRS tracking data for SKYWARN events. I had been told to look at software from *GRLevelX*. Using a combination of pay software and add-ons, it appears possible. However, I wanted a way to do it for free. And I found it!

In this section, I am going describe the process I followed to:

- Get a mapping program
- Show watch and warning data
- Show local and national radar data
- Show APRS tracking data
- All for *free!*

FIRST, GET A MAPPING PROGRAM

To start, download and install the *Google Earth* program. Once installed, launch the program. When you launch the program, you will see the screen is divided into two sections. On the left is your working panels and on the right is the large viewing area. Your left panels are divided into three sections: Search, Places, and Layers. We will be using all three of these (**Fig 1**).

Once *Google Earth* is open, we want to set some basic settings. In the Layers panel (**Fig 2**), I recommend showing the following items:

- Roads
- 3D Buildings
- Borders and Labels
- Terrain

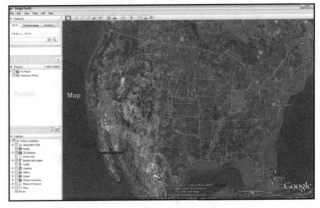

Fig 1

However, the choice is up to you to show more or less detail. Now, the observant person may see an option for Weather right there. I do not like this option because they are pulling their data from Weather.com. They also only provide a national composite radar image. This means that Weather.com needs to grab the data from the National Weather Service (NWS) and stitch it all together into a national map image. This causes both a slight delay and a loss of some precision.

Fig 2

Appendix 7 134

Now that we have the map showing the features we want, use the Search panel to get zoomed in on your county. I recommend adjusting the zoom to show your county and about one county more around it on all sides. This gives you a good view of what is coming your way. Once you have the zoom adjusted, click on the View menu and select "Make this my start location." By doing this, *Google Earth* will automatically go to this view when the program is opened saving you the need to search for your location each time (**Fig 3**).

Fig 3

ENTER KML

So at this point, we have met the first bullet of our requirements. But how can we get data directly from the NWS or APRS data into *Google Earth*? It is done through the magic known as *Keyhole Markup Language (KML)*. KML is an extension of the *eXtensible Markup Language (XML)*. KML allows for the overlay of images, data points, polygons, icons and links to inline rich-text data, HTML, or external Web sites. Often, KML files will be compressed or zipped into a KMZ file. (Note to nerds: you can download a .KMZ file, rename it .ZIP, and extract the .KML file from it. You can then rename the .KML file to .XML and open it to see the raw data.) Keep in mind that what you see may simply be a link to Web data. We will see two examples of this as we go down our shopping list of wants.

ADD WATCH AND WARNING DATA

Our next set of requirements requires data from the National Weather Service. All the data that we want to use is already provided. And now, the NWS provides this data via KML format! And to make it super easy, they created a KML / KMZ generator page that puts all items in one simple-to-use page (**Fig 4**).

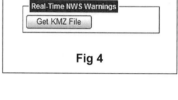

Fig 4

The second bullet is to see current watches and warnings. This is crucial to see what current hot spots exist. To add this data to *Google Earth*, go to the KMZ Generator page and click the "Get KMZ File" button for Real-Time NWS Warnings. Click the Open button when prompted. *Google Earth* will launch if it is not already open (**Fig 5**).

In the Temporary Places section of the Places panel, we see the NWS Warnings item is added (**Fig 6**). We expand this to see there are two items under NWS

Fig 5

Warnings. The first is a folder called National Weather Service. Notice the standard folder icon showing this contains static information. Inside the National Weather Service folder are three items: National Weather Service, NOAA and Warning Legend. The first two add logos to the screen while the third adds the warning legend: Red for Tornado, Yellow for Severe Thunderstorm, Green for Flash Flood, and Blue for Marine. I don't particularly like parts of my screen getting covered by the NWS and NOAA logos. To get rid of them, you can simply right click on those items and select delete.

Fig 6

The second folder under NWS Warnings is a little different. You will notice this folder has a little network line under it. This tells us that the content of this folder is not static and is actually pulled from the Web live. If we right click on the NWS Warning Products and select Properties, we can see what is really happening. Right on top, we see the link to **http://www.srh.noaa.gov/data/radar/poly.kml** We start to see the nesting capabilities of KML. The initial KML file contained static items as well as this Web link. Therefore every time we launch *Google Earth*, it will pull in the current information. (Note to nerds: you can directly put this URL into your browser, save the .KML file, rename to .XML, and open it to see the raw data.)

When we first selected this item, it was labeled "Real-Time". We can review the refresh details for this network link. We can see here that it is set to pull new data every minute. So while it is labeled as real time, we are actually doing an update every minute. If you watch your screen closely and look at that network folder icon, you will see an animated folder icon as it is updating from the Web. You will see 'data' flowing over that little network line on the bottom of the folder (**Fig 7**).

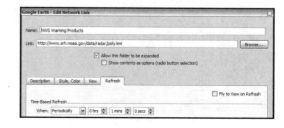

Fig 7

When we look at the map and see a warning, we can click on the little icon associated with the warning polygon. In this example, when I click on the yellow lightning bolt box on the upper right of the yellow warning area, I get a pop-up box right in *Google Earth* with the complete details of the warning, sightings, times, and locations (**Fig 8**).

Right now, our custom version of the NWS Warnings folder is stored in Temporary Places. As the name implies, when you close *Google Earth*, this will be lost. To save this, drag and drop the NWS Warnings folder from Temporary Places to My Places. Then click on the File menu, select Save, and Save My Places. Inside the My Places folder will also be your Starting Location saved from above.

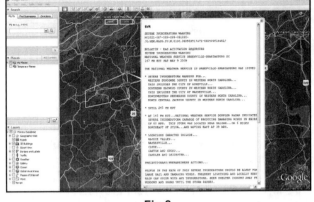

Fig 8

If you wanted to remove this, right click and delete it or just resave your starting location using the method above (**Fig 9**).

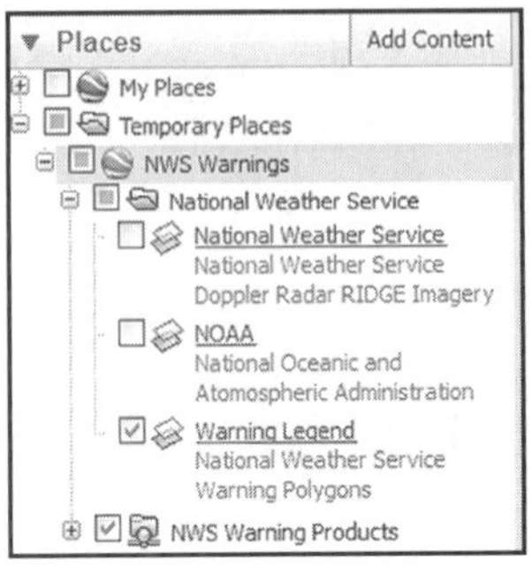

At the front of each folder and item is a check box. When you check something at a higher level, all sub-levels and items will be checked. Therefore, if you did not want to delete those logos, you could simply uncheck them. However, for the ease of just being able to click the collapsed NWS Warnings folder and have all the contents selected and shown, it is great. If you want to see what each element does, use the individual check boxes to turn each item on and off.

Fig 9

ADD RADAR DATA

On to the next bullet: radar images. For my personal preferences, I actually wanted both local and national radar data. I find the local radar image has a little more detail than the national image. From the local radar site, NWS actually provides several types of radar images. Because of this, I actually create three different radar image items. For quick access, the first one I add is the Short-Range Reflectivity from the La Crosse site (**Fig 10**).

Next, I add the All Images for the La Crosse site. This gives me a folder that contains each of the seven different images they offer. Finally, I add the Lower 48 States (CONUS) National mosaic reflectivity image (**Fig 11**).

As you may have seen before when you were deleting items, you can also rename them. So, after deleting unwanted elements, moving them from Temporary Places to My Places, reorganizing, and renaming, I get a structure that looks like the image in **Fig 12**. This allows me to quickly select which items I want to see.

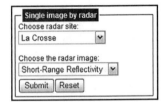

Fig 10

Fig 12

Fig 11

When I turn on the Continental US data, I can see an overlay for the entire county and where any action is. Since the CONUS image requires building the image for the entire US, it will overlay with the local radar image. I recommend using only one or the other at a time, as they may not line up exactly over each other. At the time I am creating this document, there is no activity in the local region, so my La Crosse (ARX) radar images doesn't show anything. However, in the Northeast, there is activity. **Fig 13** is a sample of that part of the country, showing the radar and warnings overlaid on the map.

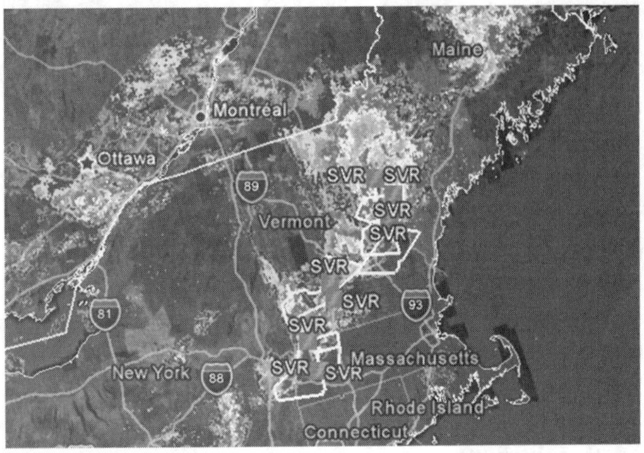

Fig 13

ADD APRS DATA

So now we have a great map tool and have integrated all the weather data we want from the NWS. The next step is to add in the APRS data. Almost as easy as it was to add in the weather data, **APRS.fi** allows us to see the APRS data stream in KML form as well. Just go to **APRS.fi** and log in with your call sign. There is no profile or account; you can put in anything you want. On the right hand side of the map, under Other Views, you will see a link for *Google Earth*. Just click that and hit Open to bring APRS data into *Google Earth*.

When this item shows up in Temporary Places, it will load all the APRS locations for the past hour in the area of the map you are looking at. This will include station locations, paths, and weather data. Again, we can notice that this is a network folder, by its icon, and not static content. This means it is talking to **APRS.fi** to get the live data. It is actually smart enough to know what region of the map you are zoomed to. As you re-zoom or move around the map, it will trigger a refresh of the data and it will pull in the APRS data for your current view. It will also refresh every five minutes if the view does not change. (Note to nerds: if you want to see how this works, right click and select Properties, as I showed above.) Due to the large volumes of APRS data, it will limit it to 1000 items on the screen. So if you pull out to show too large an area, such that greater than 1000 items would need to be drawn, it will display only some of them. A message will show on the screen indicating that the number of items was limited and to zoom in to reduce the area to see all items.

In addition to general APRS traffic from the past hour, you can track specific objects. After logging into **APRS.fi**, put the call sign and *SSID* (specific number for users that have multiple APRS devices) into the Track text box and hit search. This will show just that item on the **APRS.fi** map. However, now when you click the *Google Earth* link, in addition to the general APRS traffic folder for the last hour, you will get a second network folder for that particular call sign. It will show the last known location and path for that person, even if it was not during the past hour. If you have multiple calls you want to track, search for them on the site, add them to *Google Earth*, and move just those tracking folders from Temporary Places to My Places. As a reminder, don't forget to save your My Places. **Fig 14** shows how I normally keep my personal My Places arranged. The Starting Location does not need to be checked. It will automatically set the map to that position. By checking that box, it will only add a "Starting Location" label to your map, which may get in the way of other data. I can quickly add in my local or national radar as well as watches and warnings. Finally, I have check boxes to add in all APRS data in my view as well as track my specific items.

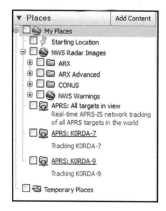

Fig 14

So that completes the list of requirements! We now have a map program, with live data from the NWS showing both warnings and radar data. We have direct access to the details on those warnings. We can see both local and national radar. And we even integrated APRS data and paths onto our map. So what does this look like when all combined? Take a look at **Fig 15**. As you can see from this sample, having all this information together makes a very powerful tool for spotters or anyone at all. And don't forget the best part: all of this is *free!*

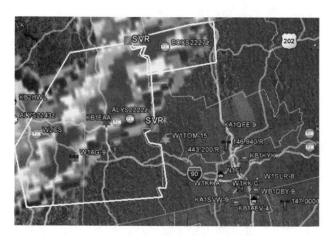

Fig 15

NWS LOOP CAPABILITIES

Recently the NWS has updated their loop capabilities. In the past, loops were hard coded to a specific time period. This means that when you added the loop, it worked for that time. However, if you came back a week later, the loop was coded to look at the original time one week ago. Now, the NWS has updated their loop code to always update in real time. This means that if I add a two-hour radar loop to my list of places, when a new 'frame' is added (every 10 minutes), it will pick it up instantly. This means that your loop is never more than two hours ten minutes old. Plus, if you save, close, and reopen, it will pull in the current time's loop. Depending on the loop you choose, it may be a one-hour or two-hour loop. The larger the coverage area, the shorter the loop and *vice versa*; smaller coverage areas give longer loops. The only thing they do not have that I would like to see is a full CONUS loop.

So how do you get these new loops? Just use the same KMZ Generator page to access the loop sections. For a single radar site, use the Single Image Loop section. For wider coverage, use the Regional Mosaic Loop section (**Fig 16**).

Because of these enhancements, I have added them into my amateur saved locations (**Fig 17**). You can see in my new panel I have two loops in addition to everything I had above. I have added in a single image loop for my closest radar site.

Storm Spotting & Amateur Radio

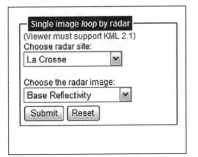

Fig 16

This gives me a two-hour loop. I also added the regional multi-radar site loop for me, the Upper Mississippi Valley. This gives me a one-hour loop. (Note to nerds: for those of you looking close, you can see how this new loop set-up works. Take a close look at the folder icons for the loops. You can see that they have the data cable icon showing that they are network folders. This means they do not store any static info but are always pulling from the Internet.) I will leave it up to you if you want to explore the properties of these new folders using the methods described above.

(Figs 1- 17: Source *Google Earth*. Used with permission.)

Fig 17

Appendix 8

Sample After Action Report

Emergency Radio Communications Drill

October 17

2009

Twenty-six City of Berkeley neighborhood groups participated in the city-wide emergency radio communications drill on October 17, 2009. The drill was intended to allow participants to practice emergency radio communications in an exercise setting and to evaluate expectations on the use of handheld FRS/GMRS two-way radios in a disaster.

After Action Report

Executive Summary

Radio communications can become an important alternate method of communication if traditional telephone and cell phone lines are disrupted. The City of Berkeley Office of Emergency Services (OES) held an emergency radio communications drill on October 17, 2009 in Berkeley, California. The scenario involved a city-wide emergency that included the loss of regular emergency communications through telephone or cell phone systems. This exercise allowed the OES to determine how effectively alternative communications to the City's Emergency Operations Center (EOC) through low-powered portable radios, through volunteer HAM radio operations, and through runners to fire stations could be implemented.

Strengths
Key strengths identified during this exercise include the following:
- Participants were able to react and adjust to shortfalls in the effectiveness of communications over the radio to the EOC.
- Messages delivered by runner to fire stations and relayed by HAM radio were effectively transmitted to the Emergency Operations Center.
- Participants developed a clear understanding of the limitations of low-powered handheld radio communications as a means of emergency communications.

Areas for Improvement
Throughout the exercise, several opportunities for improvement in the implementation of radio communications as an alternative means of emergency communications in the City of Berkeley were identified. Major recommendations include the following:
- Because firefighters will not likely be in stations in a large scale earthquake scenario plans for access to fire stations should be clarified.
- Radio channel Band Plan should be clarified.
- Expansion of exercise participation to include Fire Department station personnel should be implemented to broaden the awareness of possible emergency communications procedures for the public and for first responders.
- Improved advertising of the drill to solicit broader neighborhood and community organization participation should be implemented.
- Radio Communications education should be made available to facilitate appropriate use of equipment and proper communications procedures over the radio.
- Radio 1610AM may not have sufficient strength to reach many in-home radio receivers.

The results of this exercise should be used to frame expectations on the most effective use of alternative emergency communications needs within the City. Subsequent exercises should test specific improvements instituted as a result of this exercise and should include a focus on localized use of low-powered radios at the neighborhood level.

Chapter 1: Exercise Overview

Exercise Name: 2009 City of Berkeley Emergency Radio Communications Drill

Duration: 2 hours

Exercise Date: October 17, 2009

Sponsor: City of Berkeley Office of Emergency Services

Type of Exercise: Drill

Funding Source: N/A

Program: Community Emergency Response Training

Focus: Response

Classification: Unclassified

Scenario: Other: Earthquake

Location: Berkeley, CA

Participating Organizations:

City of Berkeley neighborhood residents

City of Berkeley Office of Emergency Services

Berkeley Fire Department

NALCO ARES

Number of Participants:

26 neighborhood representatives

2 City of Berkeley Office of Emergency Services staff

7 Northern Alameda County Amateur Radio Service (NALCO ARES) volunteer ham radio operators

7 Berkeley Fire Department Fire Station personnel

Exercise Overview:

The City of Berkeley Emergency Radio Communications Drill was a 2 hour city-wide drill designed for neighborhood Community Emergency Response Training participants. The drill was advertised via mass emails and at presentations and CERT classes leading up to the exercise. The drill was open to any participant within Berkeley who obtained a two-way FRS/GMRS radio. Participants were provided with a participants guide detailing the drill timelines and providing basic radio communications basics via email prior to the exercise. There were no pre-drill meetings and all participation was coordinated through email.

Twenty-six neighborhood groups registered and participated. Office of Emergency Services staff coordinated roll call and the exercise timeline. Northern Alameda County Amateur Radio Service (NALCO ARES) volunteers provided HAM Radio communications support from each of the 7 Berkeley fire stations. On-duty fire suppression staff tested communications from the fire stations through FRS/GMRS radios.

Exercise Evaluation:
The drill was a three part exercise. Goals included evaluating radio transmission and reception from the City's Emergency Operations Center location to locations throughout the city, familiarizing participants with their local fire station's possible role in emergency communications in a disaster and to provide an opportunity for participants to practice localized communications over the radio within their neighborhood fire districts.

A post exercise evaluation survey was utilized to obtain feedback from neighborhood participants and to analyze drill effectiveness. OES and Fire Station personnel involved in the exercise at the EOC and at Fire Stations provided exercise evaluation and observations to be integrated into the after action report.

Chapter 2: Exercise Goals and Objectives

The exercise goals established for the City of Berkeley Emergency Communications Drill were met during drill play. Through completion of the objectives, participants demonstrated effective understanding of scenario events.

The following design objectives were used in the drill:

1. Evaluation of radio transmission and reception from the City EOC. Although reception and transmission success from the City EOC was not strong, the discovery of this deficiency satisfied the evaluation objective.
2. Familiarization with local fire stations and its role in alternative emergency communications.
3. Exposure to localized neighborhood emergency communications over the radio.

Storm Spotting & Amateur Radio

Chapter 3: Exercise Event Synopsis

Scenario
A large scale earthquake or other city-wide emergency that eliminates regular emergency communications by telephone impacts the city. The scenario necessitates alternative emergency communications through low-powered hand-held radios, through local volunteer HAM radio operators at fire stations and through messenger runners. On October 17 at 9am, the City's Emergency Operations Center broadcasts on a pre-designated emergency call channel on the FRS radio band to obtain situation and status reports from neighborhoods throughout the city.

Drill Timeline

0900hrs	Welcome from "Fire Control", located at the City's Emergency Operations Center (EOC) at 2100 Martin Luther King Jr. Way "This is a drill."
0905hrs	Fire Control begins Roll Call on Channel 9 PL 9
0915 hrs	Roll Call repeated for groups that did not respond originally. Fire Control I nitiates relay messaging, if necessary.
0930hrs	Communication or delivery message to local fire station / Ham Radio operator will be at the station for receipt of message.
	Ham Radio operator transmits message to Net Control at the EOC.
1000 hrs	Participants change to a designated radio channel based on Fire District to communicate with other participants in the same Fire District.
1015hrs	Fire Control begins Roll Call for status check on Channel 9 PL 9
1030 hrs	Fire Control signs off. "End of Drill."

Chapter 4: Analysis of Mission Outcomes

OES staff was unable to copy most of the incoming radio transmissions from drill participants from the City's EOC. During the 9:05am Roll Call, only six of the twenty-six participant communications were received at Fire Control. Of those six, only two were of relative clarity; others were broken or had considerable static. Line of sight necessary for use of FRS/GMRS technology was determined to be less than effective from the EOC location.

Due to previous radio drill exercises and past practices on the assignment of radio channels, some participants expressed confusion regarding specific plans that the City has for utilizing FRS/GMRS technology in a disaster or emergency. A Radio Band Plan that is specific and that combines stability with flexibility should be developed and education on this topic disseminated.

During the exercise, a directive from Fire Control to participants to listen to Radio 1610AM was communicated via relay from the Panoramic neighborhood participant. This action was not pre-scripted and most participants successfully completed this objective. Reports from participants made it evident that the broadcast reception strength for this AM station was insufficient to many parts of the city. The output strength from this broadcast is regulated and cannot be increased and therefore, education on this limitation will be necessary to inform residents.

Conclusion

This drill was successful in furthering an understanding of the limitations of FRS/GMRS radio communications for use in emergency communications within the City of Berkeley. The exercise also served as a valuable training opportunity for participants on the basics of radio communication. Additionally, exercise participants were able to:

- Gain awareness of local fire stations and fire districts
- Gain an understanding of the function of HAM radio operators in emergency communications.
- Benefit from practice of radio operation and communication on the radio.
- Learn about the limitations regarding transmission, reception and interference inherent in emergency radio communications in a disaster.
- Connect and communicate with other participant neighborhoods in their local area.

Exercise participants met all the planned drill objectives.

Exercise participants and evaluators identified several areas for improvement in planning for emergency radio communications on the radio. Among the most critical were the clarification of expectations on use of this technology for emergency communications and the clarification of a Radio Band Plan designating call channels. These improvements are essential in coping with the inherent limitations of radio communications as an alternative in a disaster.

Improvement Plan Matrix

Task	Recommendation	Improvement Action	Responsible Party/Agency	Completion Date
Education on Radio Communications and Radio use	Radio communications protocol and equipment education would improve user proficiency.	OES is planning on adding a Radio Communications curriculum to its CERT classes	Berkeley Fire Department/ OES CERT Program	Spring, 2010
Education on Radio 1610AM as a source of Emergency Information	Radio 1610AM as a source of emergency information and the limitations of this media should be reinforced.	Provide education on Radio 1610AM to the public through CERT public safety presentations and community contacts.	Berkeley Fire Department/ OES Staff	Ongoing
Increase Participation in Future City-Wide Drills	Advanced lead time and broader publicity should be utilized to increase participation in future drills.	Targeted and broad notice to interested parties will be provided with more advanced notice for future events.	Berkeley Fire Department/ OES Staff	Ongoing
Specify Basic Plan for Emergency Communications	A Radio Band Plan, a plan for fire station access, and/or appropriate expectations should be communicated to the community.	Develop a stable basic outline of expectations to provide to the community regarding alternative emergency communications in a disaster.	Berkeley Fire Department/ OES Staff	Summer, 2010

By Ron Block, KB2UYT

Appendix 9

Lightning Protection for the Amateur Station

Part 1—Lightning protection can be a serious issue for amateurs. In the first of this three-part series, the author leads us through the process of developing a protection plan. Next: how to protect your equipment. The final part will cover the process of creating an effective ground system.

The Challenge

The amateur is challenged to assemble the best radio station possible, enjoy the benefits of the hobby, and have the station operable during times of need. This can be a significant challenge especially considering the awesome capabilities of Mother Nature's lightning strikes. While she may have the upper hand as far as when and how much energy she delivers, you have the ability to influence how that energy is diverted into the earth. Said another way, you can implement a lightning protection plan that will protect your Amateur Radio station, even from a direct strike!

The commercial radio folks have done this for years; many of them have critical installations located on hills or mountaintops that are great lightning strike targets. They do survive direct strikes and continue to provide important services to the communities that they serve. While this type of solution is possible for the Amateur Radio station, it does cost money and it does take a significant amount of resourcefulness, ingenuity, and effort to implement and maintain.

The plan does work; *but* you must follow all of the rules—exactly. Any violation of the rules, even just a little one, may result in a violation of the protection plan and damage to your equipment. In some cases the damage to a semi-protected radio station could be worse than if no protection plan had been implemented at all. I'll start with some background.

Lightning Characteristics

The conditions necessary for an old-fashioned summer afternoon thunderstorm are lots of moist air from ground level to a few thousand feet, cooler air above with little to no wind, and plenty of sun to heat the air mass near the ground. As the warm, moist air is heated, it rises quickly to heights where the temperature is below freezing, eventually forming a thundercloud as shown in Figure 1. Within the thundercloud, the constant collisions among ice particles driven by the rising air causes a static charge to build up. Eventually the static charge becomes sufficiently large to cause the electrical breakdown of the air—a lightning strike.

The average thunderstorm is approximately six miles wide and travels at approximately 25 mph. The anvil shape of the cloud is due to a combination of thermal layer (tropopause) and upper high velocity winds that cause the top of the cloud

Figure 1—A typical convective thunderhead.

to mushroom and be pushed forward. The area of imminent danger is the area up to 10 miles in front of the leading edge of the cloud.

When a lightning strike does occur, the return stroke rapidly deposits several large pulses of energy along the leader channel. That channel is heated by the energy to above 50,000°F in only a microsecond and hence has no time to expand while it is being heated, creating extremely high pressure. The high pressure channel rapidly expands into the surrounding air and compresses it. This disturbance of the air propagates outward in all directions. For the first 10 yards or so it propagates as a shock wave (faster than the speed of sound) and after that as an ordinary sound wave—the thunder we hear.

During a lightning strike your equipment is subjected to several huge impulses of energy. The majority of the energy is pulsed dc with a substantial amount of RF energy created by the fast rise time of the pulses. A typical lightning strike rise time is 1.8 µS. That translates into a radiated RF signal at 139 kHz. Rise times can vary from a very fast 0.25 µS to a very slow 12 µS, yielding an RF range from 1 MHz down to 20 kHz. However, the attachment point for a direct lightning strike has a time as fast as 10 nS. This RF content of the strike

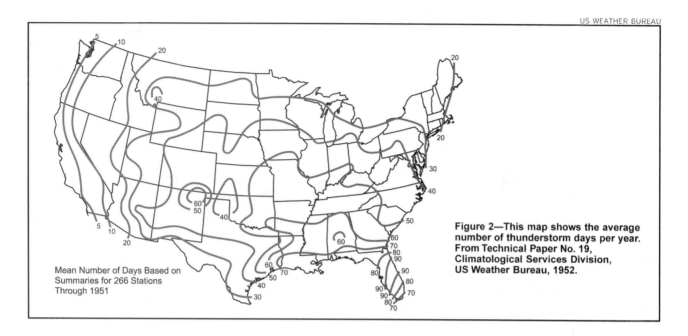

Figure 2—This map shows the average number of thunderstorm days per year. From Technical Paper No. 19, Climatological Services Division, US Weather Bureau, 1952.

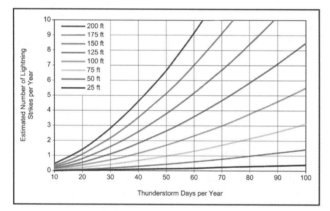

Figure 3—Estimated number of lightning strikes per year based on the number of thunderstorm days in your area and the height of your antenna. Based on information from *Living with Lightning*, Seminar Notes #ECP-826B Version F, GE Mobile Radio Technical Training, © GE 1985.

will have a major effect on the design of the protection plan. In addition to the strike pulses, the antennas and feed lines form tuned circuits that will ring when the pulses hit. This is much like striking a tuning fork in that ringing is created from the lightning's pulsed energy.

Average peak current for the first strike is approximately 18 kA (98% of the strikes fall between 3 kA to 140 kA). For the second and subsequent impulses, the current will be about half the initial peak. Yes, there is usually more than one impulse. The reason that we perceive a lightning strike to flicker is that it is composed of an average 3 to 4 impulses per lightning strike. The typical interval between impulses is approximately 50 mS.

Vulnerability

Frequently, amateurs provide an inducement for Mother Nature to find us. For good long distance communications, we put our antennas on the top of towers and place the towers so that they protrude above the surrounding buildings or countryside. While this provides for great signal coverage, it also makes it easier for Mother Nature to find a shorter, conductive path to ground.

The probability of having your tower struck by lightning is governed primarily by where you are located and the height of the tower. In 1952, The Weather Bureau compiled a contour map of the US showing the mean number of thunderstorm-days that occur in a year shown in Figure 2. The counting criterion is relatively simple—a thunderstorm-day is one in which one or more claps of thunder are heard. This gives us a reasonable view of the country with respect to our exposure to lightning.

The other significant factor that affects the probability of being struck is the height of the tower above the average ground level. As you might suspect, the higher your tower, the higher the probability of being struck. Figure 3 shows the estimated number of times per year that a tower of a given height would be struck based on the number of thunderstorm-days in your area.

Now that you can estimate your approximate exposure, you might have one of several reactions. First, you could say the predictions are all wrong—personally, I have never had my 100-foot tower struck since I put it up two years ago. Maybe you are just lucky or the law of averages has yet to catch up with you. Another reaction you could have is, "Wow! This explains what has been happening and I had better do something about this right away."

In either case, this article shows what you need to do to protect both your life and equipment, broken into straightforward steps to maximize your probability of success. Every Amateur Radio station is different and there is no "one size fits all" solution. There are, however, some well grounded (pun intended) principles that must be followed. A failure to follow the principles will result in the expenditure of both time and money with no better protection (possibly even worse damage) than if you had done nothing at all. Please follow each step carefully, thinking about the principles involved, and carefully apply the information to your specific installation.

Identify What is to be Protected

The goal of the planning process is to establish a "zone of protection" within the radio room, as opposed to the whole house or building. Additional zones may be considered separately. The first step in the process is to identify what you want to protect. The immediate answer is, well, everything. While you can come close, you may run out of money, time, or energy. So let's create a priority list and work the list from

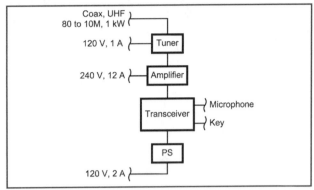

Figure 4—Block diagram of a typical simple HF radio station.

high to low priority. Probably first on the list are the more expensive items associated with your radio station, usually the transmitting and receiving equipment. Viewed another way, without these there is no radio station, so they should be very near the top of the list. What follows on the list depends on just how you enjoy the hobby—the antenna tuner, linear amplifier, terminal node controller, or computer. Further down the list might be the antenna, rotor and transmission line. Each person's list and priority ordering will be different. Pause here and mentally construct your priority list, being sure to include all the elements of your radio station. We will then work through the process of developing your protection plan.

The first step is to construct a complete block diagram of the equipment in your radio room starting with the top priority item. (You will make a separate plan for other areas needing protection.) This is usually simple and straightforward. In some installations it may be necessary to look behind the equipment to determine precisely the connections between each element. The accuracy of the diagram is important in determining the nature and effectiveness of the protection plan.

I would imagine that the list's top priority items are your transmitter and receiver (or transceiver). If you have multiples of either, then they are probably listed in order of value. These are the heart of your radio station, so make them the starting point of protection plan which will in turn examine and diagram each element of the station.

Assuming your primary item is a transceiver, represent it in the block diagram as a single rectangle. Label it with the manufacturer's name and model number. If your primary equipment is a transmitter/receiver pair, then represent them as individual single rectangles.

Next, think about the antenna connection to the primary transceiver, transmitter, or receiver. If the connection goes directly to the external antenna, simply draw a line from the rectangle to the edge of the paper. However, if the antenna is connected to the equipment via a linear amplifier, antenna tuner, or a multi-position coax switch, add this (these) as separate rectangle(s) interconnected with the primary radio equipment. The feed line going to the antenna should still go to the edge of the paper. Label the feed line's lowest and highest frequency (MHz or band name), the maximum transmit power in watts (rounded up), and the type of connector and gender (UHF/PL-259 male or N-series male, for example).

Add a rectangle to the diagram for each additional transceiver, transmitter, amplifier, and receivers in your radio room. Be sure to show all interconnections and antenna connections for each of these secondary rectangles. If any of the secondary radio equipment has a direct connection to an antenna, show the feed line going to the edge of the page. Be sure to label each rectangle with the manufacturer's name and model number and each feed line with connector type and gender, frequency range, and maximum transmit power. Figure 4 shows a block diagram for a simple station.

The block diagram should now have a rectangle representing each piece of radio equipment and accessories in the radio room. Each of the rectangles should have lines representing the interconnecting cables and feed lines. Each feed line that leaves the radio room and goes to an antenna or some tower-mounted electronics should be drawn to the edge of the page and labeled.

A Close Look

Now it is time to examine each of the rectangles, one at a time, and to add to the diagram any other electronic devices (as rectangles), complete with the electrical connections and interconnections between them. Some of these will be easy and intuitive, while others will require a little more crawling around behind the equipment. Every connection must be included—this is important to the integrity of the solution. The only exception is a non-conductive fiber-optic connection.

To complete the diagram in an orderly fashion, pick a rectangle and answer all of the following questions for that rectangle. Then, pick another rectangle and do the same until all of the rectangles have been examined.

Is there a connection between this rectangle and any other rectangle? If so, add a line between the respective rectangles and label its function.

Is there a connection between this rectangle and a device not yet included on the block diagram? This can include standalone amplifiers, power supplies, computers, terminal node controllers, modems, network routers, network hubs, and the like. If so, add the new device to the diagram as a rectangle and label it. Then add and label the connections. Repeat this step until all connections from this rectangle to new devices have been completed.

Is there an ac power connection required for this rectangle? If so, draw a line to the edge of the page and label it with the voltage and current required.

Is there a requirement to supply ac or dc power through a feed line to operate remote switches or electronics? If so, label the feed line at the edge of the page with the peak voltage and current requirements.

Are there control lines leaving the rectangle going to remote electronics, relays, or rotors? If so, draw a line to the edge of the page and label it appropriately.

Is there ac power leaving the rectangle going to the tower for safety lighting, convenience outlets, crank-up motors, or high-power rotors? If so, draw a line to the edge of the page and label it with the voltage and current required.

Is there a connection to a telephone line, ISDN telephone circuit, DSL telephone circuit, or cable connection (RF, video or data) for this rectangle? If so, draw a line to the edge of the page and label it appropriately.

Is there a connection to another antenna system such as for GPS, broadcast or cable TV, or DBS dish for this rectangle? If so, draw a line to the edge of the page and label it appropriately.

Is there a connection to other equipment elsewhere in the house or building, such or network or intercom cabling? If so, draw a line to the edge of the page and label it appropriately.

Once you have completed the process for each of the rectangles, including all of the new ones that were added, you should have an accurate block diagram of your radio station. It may be prudent to review each rectangle to verify that nothing was left out. Your block diagram should look something like Figure 5.

Now step back and physically look at the equipment in the radio room. Has every piece of equipment been reflected in

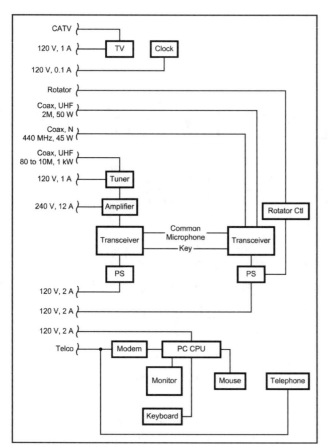

Figure 5—Block diagram of a typical more-complex radio station.

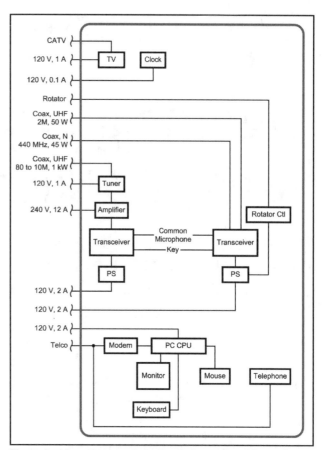

Figure 6—Lines that penetrate the circle are the radio station I/O circuits that must be protected.

the block diagram? Every metallic item within four feet (in all directions) of the radio equipment must be considered as a part of the radio equipment even if it is not electrically connected to it. If there is such an item that has not been included, we need to carefully examine it. An example of such a device could be a simple stand-alone telephone on the operating desk or a computer system (CPU, monitor, keyboard and mouse) some part of which is sitting on or near the radio desk.

Nearby devices (telephone and or computer), while electrically not a part of the radio station, are within a spark-gap of the radio station equipment and therefore considered proximally connected to the radio station and must be added to the block diagram. Follow the same procedure that you used to add equipment to the block diagram. As an example, Figure 5 also shows a computer that is included in the protection plan, but not directly connected to radio equipment.

Now that the diagram is accurate and complete, draw a circle around all of the rectangles allowing each of the lines that extend to the edge of the page to cross the circle as shown in Figure 6. The equipment represented by the rectangles within the circle is to be protected. All of the lines going from the circle to the edge of the page are called I/O (Input/Output) lines or circuits.

All or Nothing

One word of caution regarding the accuracy and attention to detail; the protection is all or nothing. If an I/O line is inadvertently missed then the protection plan is flawed and the damage could be worse than having no protection at all.

Please note: Just because equipment may survive a direct lightning strike, does not mean that you can. You *cannot* operate (touch) the equipment during a strike because *you* breach the protected equipment circle to the outside world. You are conductive, and it could hurt both you and your equipment.

Now that you have identified all of the I/O lines for the station, each must be protected and each of the I/O line protectors must be grounded and mounted in common. We will discuss how to do this in the next part of this article.

Ron Block, KB2UYT, has been a distributor and consultant for PolyPhaser, a vendor of lightning protection systems, since 1989 and has completed The Lightning Protection Course *by PolyPhaser. He is the chairman of the* Amateur Radio Station Grounding *forum at the Dayton Hamvention and has been a guest speaker at various Amateur Radio club meetings. He may be reached at 327 Barbara Dr, Clarksboro, NJ 08020; ron@wrblock.com. The author's brother, Roger, founder of PolyPhaser, reviewed this article for technical accuracy.*

Storm Spotting & Amateur Radio

By Ron Block, KB2UYT

Lightning Protection for the Amateur Radio Station

Part 2—Last month, the author discussed the characteristics of lightning and the hazard it presents to the amateur, and presented a method of preparing a schematic of a protection plan. This installment shows us the type of protection to apply and how to design the protective installation.

The process described in Part 1 of identifying the equipment to be protected can be applied to any item or set of electronic equipment. With some adjustment, it can be applied to tower-top electronics such as preamps or power amplifiers, to a computer installation in another room, a TV or a stereo system. The principle is the same: identify all of the electrically and proximally connected equipment, identify all of the I/O (input/output) lines, add protectors and ground. The theory is easy; it's the implementation that can be challenging.

Protecting Each I/O Line

Let's examine each of the I/O lines identified in the box-level schematic, dividing them into broad categories for discussion. Each I/O line represents a potential source or sink (ground) for lightning strike energy, either directly from Mother Nature or indirectly via a connecting wire or arc. We must provide a protector that is physically and electrically appropriate for the type of I/O line we are protecting. The protector has a relatively simple job to do—short circuit when threatened (over voltage). While this may seem like a relatively simple thing to do, it is surprisingly difficult to accomplish without first sharing much of the strike energy with your equipment. This is especially important for receivers with sensitive FET front-end stages and electronic interfaces (RS-232, 422, and so on) where the maximum tolerable interface voltage is just a few volts above the operating voltage.

The best I/O line protectors are connected in series between the surge and the circuit they are intended to protect. Series protectors, by design, have the capability to limit the amount of lightning strike energy your equipment will receive. The "better manufacturers" will specify the maximum amount of "let-through energy" your equipment will receive during a strike. It is normally specified as a quantity of energy in the milli- or microjoule range. When choosing a protector, select the one with the least let-through energy that meets all of the requirements for the connection.

Coaxial Cable

The first category of protector we will examine is the coaxial cable line protector. Coaxial protectors are unique in that they should not add to system SWR or signal loss, and at the same time they need to operate over a very broad frequency range at both receive and transmit power levels.

Each coax line leaving the circle around the protected equipment must have an appropriate coaxial protector. As we will discuss later, the coax protector along with all of the other I/O protectors must be mounted on a common plate (or panel) and connected to an external ground system.

Two typical PolyPhaser protectors for Amateur Radio use below 1 GHz are shown in Figure 7. While both of these protectors are shown with UHF-female connectors on both the antenna and equipment sides of the protector, type N connectors are available, as are combinations of male and female connectors. Please note, however, that there are other manufacturers of quality lightning-protection products. See the "Resources" sidebar at the end of this article.

Special coaxial protectors to protect I/O lines for GPS, DBS, broadcast and cable TV are available, as well as those for tower-top amplifiers and remote antenna switches that require an ac or dc voltage fed through the feed line. All protectors come with the appropriate type of connector commonly used for these applications.

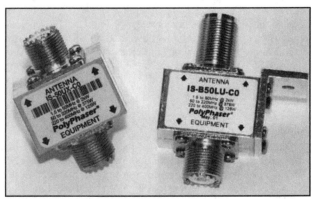

Figure 7—Typical coax protectors, the PolyPhaser IS-50UX and IS-B50LU.

Figure 8—An in-line ac power protector.

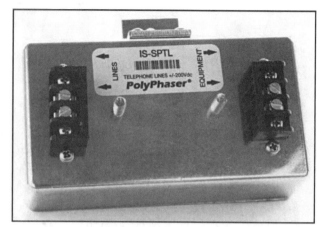

Figure 9—A telephone line protector.

Open-Wire or Ladder Line

Although protecting an open-wire or ladder line is not as convenient as with a coax line, some protection is warranted and possible. Select two identical gas tube protectors and connect them from each leg of the feed line to ground near the entrance point. Each gas tube should be specified as capable of handling an instantaneous peak current of approximately 50,000 A based on the 8/20 µS IEEE standard test waveform and have a turn-on voltage that is well above the normal transmission line operating voltage. Be sure to consider the highest SWR and the highest transmit power in your calculations. Typical turn-on voltages range from 600 to 1200 V. The voltage chosen should be about twice the calculated voltage to minimize the potential for the accidental firing of the gas tubes during tune-up or other transmitter anomalies.

Keep in mind that the application of the gas tubes to the open-wire or ladder line represents a shunt-type connection, as opposed to the coaxial protectors, which are an in-line connection. That means that the transmission line will share a significant amount of lightning strike energy with your equipment before the gas tubes begin to conduct. Unfortunately, this type of transmission line makes it difficult to achieve a high level of confidence in protecting high performance receivers.

AC Power

AC power protectors are available in many shapes, capabilities, and method of connection. Some caution should to be exercised in choosing your protector. There are many rather inexpensive power line protectors on the market that are clearly not suitable lightning protection. Many of these protectors depend on the safety ground wire to carry away the surge energy. While the safety ground may provide a dc path to ground, the #14 AWG wire commonly used is too inductive with respect to the rise time of the currents (RF energy) in the strike that it must conduct to ground. In addition, some low-end manufacturers who do provide in-line ac protectors use ferrite core inductors to maintain a small sleek physical appearance.

While this approach works well when the protection is merely handling power line noise, the inductor saturates under the massive current of a real strike and the benefit of the inductance disappears. Plastic housings and printed circuit boards should be avoided where possible since they will most likely not hold up under real strike conditions when you need it.

Since you are establishing a local zone of protection for the radio room you need to choose an in-line ac power protector, as shown in Figure 8, that matches your voltage and current requirements. For most small to medium size stations, a single 120 V ac protector with a capability of 15 or 20 A will satisfy all of our ac power needs. Each of the electronic items with an ac power line extending beyond the circle should be aggregated into a single line as long as it is comfortably within the maximum amperage of the selected protector (usually 15 or 20 A). Larger stations with high-power amplifiers or transmitter will most likely have a separate 120 V ac or 240 V ac power circuit that will require a separate ac power protector. Some high-end stations may require 100 A or 200 A in-line protectors.

If station ac is sent outside for convenience, for safety lighting, or to run motors (not the common antenna rotator), then that ac circuit must be separately protected as it leaves the radio room.

Telephone

Telephone lines come in many types, but by far the most common is the plain old telephone service (POTS). This is a balanced line with a –48 V dc battery talk circuit and up to 140 V ac ringing voltage. An in-line protector is the most effective type for POTS with different types of protectors available for different telephone line characteristics. One device for this purpose is shown in Figure 9.

A word of caution—many of the protectors on the market use modular connectors (RJ-11, -12, -45). While this is a great convenience for the installer, electrically this is a very fragile connector and common amounts of surge energy are very likely to destroy the connector by welding it or fusing it open. In addition, there are also issues of flammable plastic housings, ground wire characteristics, and printed circuit boards that allow arcs to the equipment side.

Control Circuits

Control circuits for all external devices must be protected, especially those that are tower-mounted. In the amateur station, this usually consists of an antenna rotator. Since most of the control circuitry managing the antenna rotator is relay-based (as opposed to electronic), we can use a less expensive shunt type protection device, such as that shown in Figure 10.

There are some new rotators on the market that use optical encoders and a modestly protected digital interface. These must

Figure 10—This shunt-type device is capable of protecting up to eight circuit lines with an operating voltage of up to 82 V dc.

also be protected. The method of protection will change, however, since the interface is electronic. Once the peak operating interface voltages are determined, it is relatively straightforward to choose the appropriate inline protector for the individual conductors.

Miscellaneous

Depending on the equipment in the radio room there may be additional I/O lines remaining to be covered. While I'll address a few of the more common ones, the others will probably require some special attention based on the physical conditions of the site.

Ethernet network cable connections linking the amateur station to the outside world or the computer in another room must also be protected as apart of the protection plan. For 10 and 100 Mbit UTP (unshielded twisted pair) networks, the use of an ITW LINX protector for Cat5-LAN (four pair) cable is recommended. This protector is wired in series with the network using 110-type punch-down blocks and grounded similarly to other protectors.

For those radio rooms that have broadcast or cable TV, protection is similar to the coaxial protectors described above with the exception that the impedance of the unit is 75 Ω and F-type connectors are used.

For single and dual-LNB DBS dishes the protector is required to have a very broad band-pass and pass dc through the coax center conductor.

GPS feed lines also are commonly required to carry a dc voltage. A high quality protector will separate the RF from the dc and protect each to its own voltage and power specification.

I/O Wrap-up

Every line that penetrates the circle and goes to the edge of the page should now have an identified protector. If you are having trouble, identify the problem area and mail the drawing to the author. The ARRL staff has compiled a list of potential sources of lightning-protection products. See the "Resources" sidebar at the end of this article.

Single Point Ground

The next step in the process will take us away from the theoretical work that we have been doing and into the real world of practical design and component layout. It's not hard, but there are a lot of things to consider as we take each step. Most of the considerations will be unique to the physical circumstances associated with your radio room.

I mentioned earlier that the primary purpose of the protector is relatively simple—to short-circuit when threatened. By shorting all of the wires associated with an interface no current can flow through the equipment between the wires of the interface. Extending this premise further, by mounting all of the protectors in common, no current will flow between the I/O interfaces. Hence, no lightning surge current will flow *through* a protected piece of electronic equipment.

To make this possible in the radio room it is necessary to establish what is known as a "Single Point Ground." This is the *one and only point* in the radio room where a ground connection is present. We need to be a little careful with the term "ground." During a strike a ground can be anything that is capable of being an energy sink. By this definition absolutely anything that is not at the same electrical potential can be a sink. Because electrical signals travel at about 1 nanosecond per foot, fast rise times may create significant potential differences for short times due to travel differences.

The creation of a single point ground will be different for every installation. It can be as simple as a couple of protectors bolted together or a through-wall entrance panel, or as complex as a copper-covered wall upon which the protectors are mounted. Whatever form your single point ground takes it must be the only ground point for all of the equipment within the circle of the box-level schematic diagram.

Figure 11 shows a single-point ground panel. This is a high-density fiberboard-backed copper panel suitable for small to medium radio rooms. It comes with a 1½ inch wide copper strap to connect the panel to the external ground system and a second 1½ inch copper strap to connect to all your operating table equipment. The panel is intended to be mounted on a wall near the radio equipment. For convenience of reference, I'll use its abbreviation—SPGP, for Single Point Ground Panel.

Now that you have a mounting surface that will become the single-point ground, a *lot* of consideration must be given to the physical placement of the protectors on the SPGP. Remember that a protector is required for each I/O line that leaves the circle of the box-level schematic. As you examine a protector, most are labeled with respect to which connector faces the surge (the outside world) and which connector faces your equipment. This is important, since the protectors are not necessarily symmetric in their design. They cannot be reversed and be expected to function properly.

A significant factor in the layout of the protectors on the SPGP is maintaining a physical separation between the incoming unprotected cables (antenna feed lines, incoming ac power, rotator lines, etc) and the protected side of the same connections. As a result of going through an in-line protector, there will be a "spark-gap level" voltage difference for a short time between

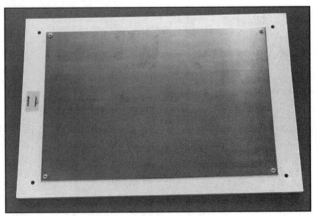

Figure 11—A typical single-point ground panel.

the input and output sides of the protector. You must take this into consideration when planning the layout of the SPGP.

A general guideline is to draw an imaginary diagonal line near the center of the panel as shown in Figure 12. Designate the area above the line as protected and the area below the line as unprotected (or vice versa). Make sure you consider how the panel will be mounted; how the (unprotected) cables will enter the unprotected area and how the (protected) cables will leave the panel. One of the nice things about a two-dimensional drawing is that the effects of gravity do not show. In Figure 12, the cables leaving the panel to the right above the dotted line must be anchored. If they are not, real world gravity will cause them to eventually bend down and come close to, and maybe even touch, the unprotected cables. If this happens, during the strike event there is the potential for a spark-gap breach of the protectors between the cables—a failure of the protection plan.

Neatness counts—cables (transmission lines, power (ac and dc), speaker, microphone, computer, control) should be cut to length and routed neatly and cleanly between boxes using the most direct practical route. The coiling of excess cable length on the protected side should be avoided since it can act as an air-wound transformer coupling magnetic energy from a nearby lightning strike back into the protected equipment.

The chassis ground for each element of radio equipment must also be connected to the SPGP. The SPGP is our reference point during the strike event and it is important that all elements of the radio station be at the same potential at the same time (nanoseconds). For small to medium size stations, where all for the equipment fits on a desk/table top, a single interconnect copper bus or strap to the SPGP is usually sufficient.

For stations with freestanding cabinets or racks in addition to an operating desk, the issue of rise time becomes more significant due to distance. This necessitates separate cabinet/rack direct ground connections to the SPGP. In addition, stations of this size have other special considerations, such as concrete floor conductivity, that are not covered here.

Don't forget to allow for future growth of your station in the SPGP layout. Typically this means leaving room for an additional feed line protector or two and maybe a rotator control protector. It is easier to plan for expansion now, rather than have to rearrange the protectors on the panel later.

If the form of SPGP you have chosen is a metal plate mounted in a window or a full-fledged through-wall entrance panel, you can ignore the remainder of this paragraph. The next major consideration is the placement of the SPGP with relation to the radio equipment. The SPGP is ideally mounted on the inside of an exterior wall with access to an earth ground and within a few feet of the radio equipment. That sounds easy, but depending on your radio room, it may be next to impossible. Let's work it through.

These real-world constraints sometimes present real challenges. One of the biggest challenges is grounding the SPGP. A #6 AWG wire to a radiator or water pipe is usually *not* acceptable! I say "usually" because if your radio room is on the top of a high-rise building, that may be all that you have. I'll discuss the real requirements and address this type of problem later.

The purpose of the ground connection is to take the energy

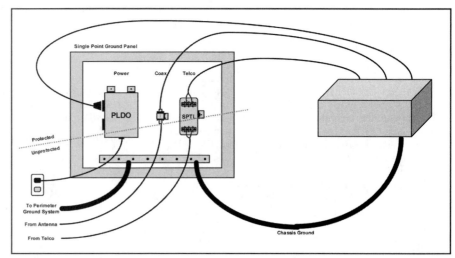

Figure 12—The SPGP showing the division of the protected and unprotected cables.

arriving on the antenna feed line cables and control lines (and to a lesser extent on the power and telephone lines) and give it a path back to the earth, our energy sink. The impedance of the ground connection should be low so the energy prefers this path and is dispersed harmlessly. To achieve a low impedance the ground connection needs to be short (distance), straight, and wide.

Short

We all know that a conductor, no matter what size or shape, has inductance that increases with length. Connecting the SPGP to the external ground system should be done with the shortest possible wire. Did I say wire? Be sure to read about "wide."

Straight

Rarely is it possible, in the context of an Amateur Radio station (unless the structure was designed around the radio station), to go directly from the SPGP to the external ground system in a short, straight line. Most of the time we are encumbered with an existing structure that is less than ideal and further encumbered with esthetic constraints regarding just how much of a mess we can make. So, we do the best we can. Straight becomes a relative concept. Run the ground wire (there's that word again) as straight as possible. Keep in mind that every time the wire makes a turn, the inductance of the path is increased a small amount; ~0.15 µH for a 90-degree turn in less than 1 inch. The cumulative effect of several turns could be meaningful. By the nature of its current (magnetic) fields, a wide wire (strap) has lower inductance per length, compared to round conductors, and has minimal inductance for turns.

Also keep in mind that speeding electric fields don't like to change direction. The inductance in each bend or turn represents a speed bump, causing a large change in the fields over a short distance. If the change is large enough, some of the electrons are likely to leave the wire and find another path to ground; that is, an arc. This is not desirable; we have lost control.

Wide

We all know that no matter what size, wire has inductance. Larger wire sizes have less inductance than the smaller sizes. We also know that RF energy travels near the surface of a wire as opposed to within the central core of the wire (skin effect). If we put these together and extend the hypothesis a little, it would seem reasonable to use a railroad rail-sized bus bar as an excellent connector between the SPGP and the earth ground.

While the large bus bar would work well, it has lots of surface area and a massive core, the cost would be prohibitively expensive and it would be extremely cumbersome to work. We can have the benefits of the large bus at a very reasonable cost if we use multi-inch-wide copper strap instead, however.

One and a half inch wide, #26 AWG (0.0159 inch) copper strap has less inductance than #4/0 AWG wire, not to mention that it is less expensive and much easier to work. We can use thin copper strap to conduct lightning surge energy safely because the energy pulse is of very short duration and the cross-sectional area of this strap is larger than #6 AWG wire. The strap has a large surface area that makes it ideal for conducting the strike's RF energy.

The goal is to make the ground path leading away from the SPGP more desirable than any other path. In order to achieve this we need to find the total amount of coax surface area coming to the SPGP from the antennas. The circumference of a single 9913 coaxial cable represents about 1.27 inches of incoming conductor surface. To make our ground path appealing to the surge energy, we ideally need more than 1.27 inches of conductor surface leaving the SPGP. Where the use of a single 1½ inch wide conductor leaving the panel is reasonable, a strap three or more inches wide would be better. Inductance is calculated on the length of the connection between the SPGP and the ground, as well as the number and sharpness of the turns. If you had three ⅞-inch Hardlines, a minimum strap width of 9 inches would be needed and 12 would be better.

You now have determined what protective devices are needed and how to mount them for an effective barrier to lightning energy. Next month, the final part of the article will present guidelines for developing a good external ground to absorb and dissipate the lightning's energy.

Photos of various PolyPhaser products by the author.

Ron Block, KB2UYT, has been a distributor and consultant for PolyPhaser, a vendor of lightning protection systems, since 1989 and has completed The Lightning Protection Course *by PolyPhaser. He is the chairman of the* Amateur Radio Station Grounding *forum at the Dayton Hamvention and has been a guest speaker at various Amateur Radio club meetings. The author's brother, Roger, founder of PolyPhaser, reviewed this article for technical accuracy.*

327 Barbara Dr,
Clarksboro, NJ 08020
ron@wrblock.com

Storm Spotting & Amateur Radio

By Ron Block, KB2UYT

Lightning Protection for the Amateur Radio Station

Part 3—In this final installment, the author shows how to develop a good external ground system to complete your station's protection.

Establishing a Good Ground

Now that the SPGP (Single-Point Ground Panel) is connected through the wall to the outside world, there is still a lot of work to do. It's necessary to switch from brainpower and the challenge of getting copper strap through walls to the brute force required to establish a good ground system. The operative word here is *system*—not a ground rod, but a network of interconnected ground rods.

The primary purpose of the external ground system is to disperse as much of the lightning energy as possible into the earth before it follows the feed line into the radio station. No matter how hard one tries, some of it will follow the coax, which is why you created the protection plan for the radio equipment. The easier you make it for the strike energy to dissipate in the earth before it gets to the radio station, the less your equipment protection plan will be stressed.

With great diligence, hard work, no real estate restrictions, plenty of funds and highly conductive soil, it is possible for up to 90% of the strike energy to be dissipated in the earth, leaving just 10% heading for your equipment. This would be quite an accomplishment. In many commercial sites it doesn't work out that well and rarely, if ever, for the Amateur Radio station—there are always restrictions. Let's see what should be done and then adjust to the home environment's restrictions.

Figure 13 shows what has to be done. In the center, the concentric triangles represent the tower. Ideally, the tower is separated from the house by 20 to 50 feet. This distance provides sufficient room for the dissipation of the magnetic fields during the strike event. This distance also takes advantage of the natural inductance of the antenna feed lines to limit the amount of surge and allow more time for the tower grounding system to absorb the strike energy.

Radials and Ground Rods

Spreading out from the base of the tower is a set of eight radials. While the number of radials required for a particular installation will be dependent on the soil conditions in your location, the system shown here is a reasonable start. Each radial is a bare copper wire (preferably, strap) buried 6 to 18 inches below grade. The radials should be positioned so that the energy is dissipated away from the house.

Connected to the radials are ground rods. The ground rods are spaced approximately twice the length of a ground rod. For an 8-foot rod, the spacing would be 16 feet. During the strike, each ground rod has a cylindrically shaped region of influence centered on the ground rod. This is the region in which the ground rod disperses the strike energy. If the rods are placed closer, the regions of influence begin to overlap and the ground rod's ability to disperse energy is diminished. Although this overlapping does not harm the ground system,

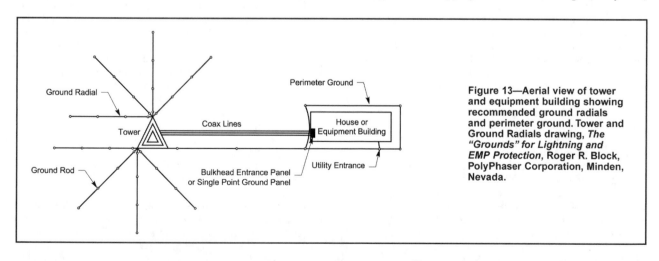

Figure 13—Aerial view of tower and equipment building showing recommended ground radials and perimeter ground. Tower and Ground Radials drawing, *The "Grounds" for Lightning and EMP Protection*, Roger R. Block, PolyPhaser Corporation, Minden, Nevada.

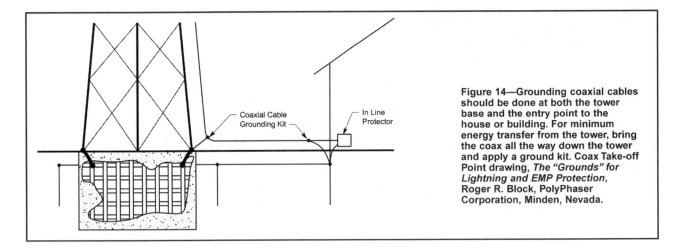

Figure 14—Grounding coaxial cables should be done at both the tower base and the entry point to the house or building. For minimum energy transfer from the tower, bring the coax all the way down the tower and apply a ground kit. Coax Take-off Point drawing, *The "Grounds" for Lightning and EMP Protection*, Roger R. Block, PolyPhaser Corporation, Minden, Nevada.

it does increase the cost since more rods are used.

Ideally, the connection between a radial and its ground rod should be made using an exothermic bonding process. This connection will most likely outlast the life expectancy of the ground system and you won't have to do annual inspections. A number of manufacturers supply the molds and fusing material for a variety of cable/strap and ground rod sizes. Two of them are Erico Incorporated (www.erico.com) and Alltec Corporation (www.allteccorp.com). If the exothermic process is not used, mechanical clamps that can be used to connect the radial to the ground rod are available. All mechanical connections must be inspected annually or more frequently to ensure the integrity of the system.

Ground rods must be mechanically driven into the ground. This is the only way to ensure that the rod achieves a reasonable "connection" to the earth. Drilling a hole and then backfilling the space around the rod is not acceptable.

While a very long radial/ground rod system is good, electrical and economic considerations come into play. Analyzing the average cost of installing an additional foot of the grounding system versus the benefit of a lower impedance system, the break-even point is somewhere around 80 feet. For practical purposes in areas with reasonably conductive soil, the maximum length of a radial should be limited to approximately 80 feet. If the impedance of the ground system needs to be lower, additional radials should be used as opposed to longer radials.

Some caution must be exercised when laying out the radials. If a radial comes within 4 feet of a metal object, it must be bonded to the metal object. The 4-foot rule applies to objects that are above, below, to the left or to the right of the radial. This includes metal fence posts, kids' swing sets, buried fuel tanks, and so on. Be sure to watch out for dissimilar metal properties when bonding to these objects.

Care must be exercised when connecting the radials to the tower. Most towers are zinc-coated steel (galvanized). Connecting a copper wire or strap directly to the tower leg will cause the zinc to erode, allowing the base steel to oxidize (rust). This in turn will increase the resistance of the connection and over time may threaten the mechanical strength of the tower segment. One solution to this problem is to use a buffer layer of stainless steel between the zinc and the copper. Several manufacturers offer tower leg clamps.

If you are constructing a new tower you can use the tower base as a "ground rod." Called a Ufer ground, it utilizes the rebar that reinforces the concrete base as an excellent ground connection. The rebar itself must be electrically interconnected so there are no spark gaps and there must be at least 4 inches of concrete between the rebar and the surrounding earth. If this is done, a wire can be brought out from the rebar and attached to the tower leg. A great big ground rod! No, you will not blow up your concrete—the other radials with ground rods will handle most of the strike energy. Since you must put rebar in the concrete anyway, use it to augment your ground system.

There are two more items that need to be highlighted in Figure 13. The first is that the SPGP (or bulkhead entrance panel) is connected to the tower ground radial system. This connection should use the same material and ground rods as is used for the radials via a buried path. If the distance between the tower and the SPGP ground point is more than approximately 100 feet, however, it may not be cost effective to interconnect the tower ground system with the SPGP ground system. In this case some portion of the tower ground system must be duplicated for the SPGP ground system.

The second item is the perimeter ground, shown in Figure 13 as going completely around the house or equipment building. This perimeter ground serves two purposes: first, it helps conduct the surge energy around the house, minimizing the ground potential differences under the house during the strike event; and second, it enhances the basic ground system by providing more connection points to the earth. The existing utility ground is also connected to the perimeter ground—this is a must!

Some Specifics

That's how it should be done; here are some general guidelines for adapting this to your specific situation.

• In general, doubling the number of radials lowers the impedance of the ground system by one half.

• Radials don't have to go in a straight line; they can follow the contour of your property or flow around obstacles. Make turns gradually (12-inch radius or larger).

• A perimeter ground that only goes three-quarters or halfway around the house is better than no perimeter ground at all. Although flowerbeds, walkways and driveways frequently present insurmountable obstacles, do your best to get most of the way around. The perimeter ground must at least connect to the utility ground.

• Short ground rods are better than none at all. Just place them closer together spaced at twice their length.

• Soil doping can improve soil conductivity. Be aware that some additives may cause ground and water pollution and can shorten the life of the grounding materials.

• Where possible when installing a new tower, place it at least 25 feet from the house; 30 feet is even better. By placing the tower at some distance from the house, you minimize the amount of magnetic energy that will couple from it into the wiring of the house. In addition, you take advantage of the inductance of the coax line in limiting the surge energy headed toward your equipment. Don't get too carried away with the distance; the added feed line also attenuates your signals.

- If you are fortunate enough to have multiple towers, each should have its own ground system of radials and ground rods. If the towers are within 100 feet of each other, a radial should be used to interconnect the towers.

Coaxial Cable Grounding

Each coaxial cable traversing your tower needs to be properly grounded to the tower. The first point is at the top of the tower where the coax connects to the antenna. The second point is where the coax leaves the tower to go to the radio equipment. This take-off point should be as close to the base of the tower as physically possible. The third point, for towers taller than 150 feet, is every 75 feet down the tower as measured from the top.

Depending on the height of the tower (inductance) and the severity of the lightning strike (current), the tower could easily have an instantaneous voltage difference between the top and the ground that exceeds 100 kV. In simplified terms, if the tower is viewed as if it were a very long resistor with one end connected to ground and the coax take-off point as a tap point on that resistor, you can begin to appreciate the problem associated with allowing the coax to leave the tower at any point above the bottom. For a 100-foot tower with a coax take-off point at the 10-foot level, approximately 10% of the tower energy (10 kV in this example) will follow the coax to your equipment. If you lower the connection to the 1 or 2-foot level, the energy flow will be correspondingly lower.

The grounding of the coax to the tower should be accomplished using a commercial-type grounding kit followed by the application of a commercial grade waterproofing material. Be careful of dissimilar metals when connecting the grounding kit pigtail to the tower.

In addition, another grounding kit should be applied to the coax just before it enters the house. This will remove some additional energy from the shield of the coax and further minimize the stress placed on your protection system. See Figure 14.

Special Considerations

Numerous Amateur Radio stations will be in locations that make it impractical or impossible to follow some or all of the advice given so far in this article. For those stations there are some things that you can do to achieve a reasonable level of protection for your equipment.

First, establish an SPGP for your equipment by creating a box-level schematic of your radio station. Identify and procure the appropriate protectors for all of the I/O connections to the radio station. Mount them on a common conductive surface. Where your installation varies from the ideal may be that the SPGP cannot be connected to an external ground.

As a substitute for an external ground, locate an alternative conductive path. Here are some recommendations for potential sources. These are in the order of most to least desirable.

- Building steel
- Stand pipe
- Metal cold water pipe
- Metal building skin
- Electrical system safety ground

Even though many of these ground choices are highly inductive and will not function as a good RF ground, the goal is to survive a lightning strike event by ensuring no current flows on the radio equipment I/O connectors. This is achieved through SPGP grounding techniques, which maintain an equal potential between equipment chassis during the strike independently of how the SPGP is grounded. For small upper-floor radio stations, this will work just fine.

Operating Safety

No matter how good your installed lightning protection plan is, you *cannot* be in electrical contact with the radio equipment during a lightning strike event. Although there is no current flowing between the radios in your radio room, all of the equipment will be statically elevated above ground. If you are holding onto the microphone or the key during the strike event, *you* are now the path of least impedance to ground from the protected and now elevated equipment chassis. This ground path can be to the rebar in the concrete floor below your feet or to a nearby electrical wire or water pipe.

Consider getting a storm warning device capable of sounding a warning when lightning activity is within 10 miles of your radio station. (Lightning detectors are available commercially, or you can build one; see Radmore, "A Lightning Detector for the Shack," Apr 2002 *QST*, pp 59-61.) When the alarm sounds, leave the radio room. If your protection plan is installed correctly, you may leave the equipment connected and powered-on—but you must leave the room. Commercial radio operators get away with operating during lightning strikes because most of their equipment is remotely operated—just like a repeater.

More Information

After you have implemented the guidelines presented here and if you have questions (recognizing that every site is unique) feel free to contact me. Be aware, the first thing that I will ask you to share with me is your box-level schematic and some physical characteristics of your radio room and antenna farm. With this information as a reference, we can discuss your situation. Please forward a copy of the appropriate information before contacting me. Some additional information is also available on the WR Block & Associates Web site at **www.wrblock.com**.

Thanks

Many thanks go to Roger R. Block, my brother and founder of PolyPhaser Corporation, for his input in the preparation of this article.

Ron Block, KB2UYT, has been a PolyPhaser distributor and consultant since 1989 and has completed The Lightning Protection Course *by PolyPhaser. He is the chairman of the Amateur Radio Station Grounding forum at the Dayton Hamvention and has been a guest speaker at various Amateur Radio club meetings. The author's brother, Roger, founder of PolyPhaser, reviewed this article for technical accuracy.*

References

Block, R. R., *The "Grounds" for Lightning and EMP Protection*, Second Edition, PolyPhaser Corporation, 1993.
Rand, K. A., *Lightning Protection and Grounding Solutions for Communications Sites*, PolyPhaser Corporation, 2000.
Uman, M. A., *All About Lightning*, Dover Publications, Inc, New York, 1986.
Uman, M. A., *Lightning*, Dover Publications, Inc, New York, 1984.
Uman, M. A., *The Lightning Discharge*, Dover Publications, Inc, New York, 2001.

[The resources sidebar that appears in Part 2 of this series (July 2002 QST, p 52) was not meant to be a complete list of manufacturers and suppliers of lightning detection equipment. To access the ARRL *TISFind* database, see **www.arrl.org/tis/tisfind.html**, and search for the key words "lightning" and "ground."—*Ed.*]

327 Barbara Dr
Clarksboro, NJ 08020
ron@wrblock.com

Index

AccuPOP 38-39
AccuWeather 38
After Action Reports 110-114
Allison House 44
Amateur Radio Emergency Service (ARES) 5, 33, 54, 103
antennas 22
APRS 9, 22, 54
ARRL Emergency Communication Handbook, The 53, 105, 108, 114
ARRL 49, 53, 107
avalanche 73
binoculars 33
blizzard 71
buffer zone 14
cell phones 27, 88
Citizen Weather Observer Program (CWOP) 45, 102
Community Emergency Response Team (CERT) .. 48, 50
damage reports 107
Debriefing Diary 114
derecho 61
dew point 55
digital cameras 29
District Emergency Coordinator (DEC) . 23, 24, 53, 104, 107
Doppler radar (see radar)
downburst 18
downdraft 63
dry adiabatic lapse rate 55
D-Star 9, 23, 24-25, 97
Echolink 9, 23, 54, 97

Emergency Communications Course 49
Emergency Coordinator (EC) .. 5, 53, 54, 90, 91, 94, 104, 107, 108, 110
Emergency Management Courses 49
Emergency Operations Center (EOC) 16, 86
Emergency Response Plan (ERP) 94
equipment 20
Espotter 38
Federal Emergency Management Agency (FEMA) 49, 104
flooding, flash flood 18, 19, 60
fog 72
FSD forms 107
GOES satellite 68, 72
go-kit 5, 10, 11, 20, 32, 33, 86, 88, 93
GPS 24, 94
GRLevelX 43
hail 17, 18, 60
hazardous weather outlook (HWO) 93, 95-96
home weather station 45, 46
hook echo 99
Hurricane Watch Net 81
hurricanes 76-91
Hurricane Debby 85
   H~ Georges 84
   H~ Gustav 86, 112
   H~ Hortense 82
   H~ Katrina 89
   H~ Lenny 85
   H~ Luis 82
Instant Messages (IM) 27

Integrated Warning System (IWS) 100
interactive NWS (iNWS) 38
Interwarn 44
jet stream 69
JetStream Online School for Weather 51
landslide 73
latent heat of condensation 55
level of free convection (LFC) 56
lifting condensation level (LCL) 56
lighting 28
lightning 16, 18, 59
Local Storm Report (LSR) 102
low level jet 58
macroburst 61
maps 26
mesocyclone 58
meteorology 55
microburst 61
moist adiabatic lapse rate 55
multi-cell storms 58
National Disaster Warning System (NADWARN) 7
National Incident Management System (NIMS) 104, 112
National Warning System (NAWAS) 94, 101
National Weather Service (NWS) 6, 17, 18, 33, 51
networking 54
NOAA Weather Radio 9, 17, 21, 94, 98
nor'easter 71
NWS Weather Safety Information 17, 18
radar 60, 65, 99
Radio Amateur Civil Emergency Service (RACES)7, 33, 54
reportable criteria 79
resources 20
safety 13-19
Saffir-Simpson Hurricane Scale 77
saturation 55
Section Emergency Coordinator (SEC) .. 23, 53, 54, 104, 107, 110
Simulated Emergency Test (SET) 53
single-cell storm 57
Skype 24
SKYWARN 5, 9, 33, 34-35, 48, 94, 97, 103, 109
sleet 71
snow 70
Specific Area Message Encoding (SAME message) ... 22
Spotter Network 41
Spotter Safety Officer (SSO) 19
squall line 58
storm chaser, chasing 11, 12
Storm Data (SD) 103
storm kit 93
Storm Prediction Center (SPC) 36, 102
StormLab 44
Stormpulse 39
stress, stress management 104-107
supercell 13, 58
The Weather Channel (TWC) 41
thunderstorms 55
TIROS I 67
tornado 13, 18, 63, 80
training 48
transceivers 20
tropical cyclone 77
tropical cyclone-induced tornado 80
tropical depression 77
tropical disturbance 77
tropical storm 77
tsunami 74
United States Weather Bureau (USWB) 65
video cameras 32
volcanic activity 73
volunteers 104-107
War Emergency Radio Service (WERS) 8
watches & warnings 37
waterspout 65
Weather Defender 43
weather satellite 67
Weather Underground 40
WeatherBug 39
WeatherTAP 42
wildfire 73
wind shear 62
wind stress 78
WX4NHC 81

F E E D B A C K

We're interested in hearing your comments on this book and what you'd like to see in future editions. Please email comments to us at **pubsfdbk@arrl.org**, including your name, call sign, email address and the title, edition and printing of this book. Or you can copy this form, fill it out, and mail to ARRL, 225 Main St, Newington, CT 06111-1494.

Please check the box that best answers these questions:

How well did this book prepare you for your exam? ☐ Very well ☐ Fairly well ☐ Not very well
How well did this book teach you about ham radio? ☐ Very well ☐ Fairly well ☐ Not very well
Which exam did you take (or will you be taking)? ☐ Technician ☐ Technician with code Did you pass? ☐ Yes ☐ No
If you checked Technician: Do you expect to learn Morse code at some point? ☐ Yes ☐ No
Which operating modes do you plan to use first?
   ☐ SSB ☐ FM ☐ Packet ☐ RTTY ☐ Image ☐ Morse code ☐ Other _____
Where did you purchase this book? ☐ From ARRL directly ☐ From an ARRL dealer
Is there a dealer who carries ARRL publications within:
   ☐ 5 miles ☐ 15 miles ☐ 30 miles of your location? ☐ Not sure.

Name _____ ARRL member? ☐ Yes ☐ No
_____ Call Sign _____

Address _____

City, State/Province, ZIP/Postal Code _____

Daytime Phone (   ) _____ Age _____

If licensed, how long? _____

Other hobbies _____ E-mail _____

Occupation _____

For ARRL use only	STORM
Edition	1 2 3 4 5 6 7 8 9 10 11 12
Printing	1 2 3 4 5 6 7 8 9 10 11 12